Extra-Solar Planets

Scottish Graduate Series

Extra-Solar Planets
The Detection, Formation, Evolution and Dynamics of Planetary Systems

B A Steves
Glasgow Caledonian University, Scotland, UK

M Hendry
University of Glasgow, Scotland, UK

A C Cameron
St. Andrews University, Scotland, UK

CRC Press
Taylor & Francis Group
Boca Raton London New York

CRC Press is an imprint of the
Taylor & Francis Group, an **informa** business
A TAYLOR & FRANCIS BOOK

CRC Press
Taylor & Francis Group
6000 Broken Sound Parkway NW, Suite 300
Boca Raton, FL 33487-2742

First issued in paperback 2019

© by Taylor & Francis Group, LLC
CRC Press is an imprint of Taylor & Francis Group, an Informa business

No claim to original U.S. Government works

ISBN-13: 978-1-4200-8344-6 (hbk)
ISBN-13: 978-0-367-38323-7 (pbk)

Visit the Taylor & Francis Web site at
http://www.taylorandfrancis.com

and the CRC Press Web site at
http://www.crcpress.com

SUSSP Schools

1	1960	Dispersion Relations
2	1961	Fluctuation, Relaxation and Resonance in Magnetic Systems
3	1962	Polarons and Excitons
4	1963	Strong Interactions and High Energy Physics
5	1964	Nuclear Structure and Electromagnetic Interactions
6	1965	Phonons in Perfect and Imperfect Lattices
7	1966	Particle Interactions at High Energy
8	1967	Methods in Solid State and Superfluid Theory
9	1968	Physics of Hot Plasmas
10	1969	Quantum Optics
11	1970	Hadronic Interactions of Photons and Electrons
12	1971	Atoms and Molecules in Astrophysics
13	1972	Properties of Amorphous Semiconductors
14	1973	Phenomenology of Particles at High Energy
15	1974	The Helium Liquids
16	1975	Non-linear Optics
17	1976	Fundamentals of Quark Models
18	1977	Nuclear Structure Physics
19	1978	Metal Non-metal Transitions in Disordered Solids
20	1979	Laser-Plasma Interactions: 1
21	1980	Gauge Theories and Experiments at High Energy
22	1981	Magnetism in Solids
23	1982	Laser-Plasma Interactions: 2
24	1982	Lasers: Physics, Systems and Techniques
25	1983	Quantitative Electron Microscopy

/continued

SUSSP Schools (continued)

26	1983	Statistical and Particle Physics
27	1984	Fundamental Forces
28	1985	Superstrings and Supergravity
29	1985	Laser-Plasma Interactions: 3
30	1985	Synchrotron Radiation
31	1986	Localisation and Interaction
32	1987	Computational Physics
33	1987	Astrophysical Plasma Spectroscopy
34	1988	Optical Computing
35	1988	Laser-Plasma Interactions: 4
36	1989	Physics of the Early Universe
37	1990	Pattern Recognition and Image Processing
38	1991	Physics of Nanostructures
39	1991	High Temperature Superconductivity
40	1992	Quantitative Microbeam Analysis
41	1992	Spatial Complexity in Optical Systems
42	1993	High Energy Phenomenology
43	1994	Determination of Geophysical Parameters from Space
44	1994	Simple Quantum Systems
45	1994	Laser-Plasma Interactions 5: Inertial Confinement Fusion
46	1995	General Relativity
47	1995	Laser Sources and Applications
48	1996	Generation and Application of High Power Microwaves
49	1997	Physical Processes in the Coastal Zone
50	1998	Semiconductor Quantum Optoelectronics

/continued

SUSSP Schools (continued)

51 1998 Muon Science

52 1998 Advances in Lasers and Applications

53 1999 Soft and Fragile Matter

54 2000 The Restless Universe

55 2001 Heavy Flavour Physics

56 2002 Ultrafast Photonics

57 2003 LHC Phenomenology

58 2004 Hadron Physics

59 2004 Soft Condensed Matter Physics in Molecular and Cell Biology

60 2005 Laser-Plasma Interactions: 6

61 2006 Neutrino Physics

62 2007 Extra-Solar Planets

Lecturers

Andrew Collier Cameron	University of St Andrews, UK
Martin Dominik	University of St Andrews, UK
Rudolf Dvorak	University of Vienna, Austria
Sylvio Ferraz-Mello	Universidade de Sao Paulo, Brasil
Eric Ford	Harvard-Smithsonian Center for Astrophysics, USA
J S Greaves	University of St. Andrews, UK
Massimiliano Guzzo	Università degli Studi di Padova, Italy
Hugh Jones	University of Hertfordshire, UK
Maciej Konacki	Nicolaus Copernicus Astronomical Centre, Poland
Andrzej J Maciejewski	University of Zielona Gora, Poland
Christian Marchal	Onera, French Aerospace lab, France
P J Message	University of Liverpool, UK
Daphne Stam	Technical University, Delft, the Netherlands
Kleomenis Tsiganis	University of Thessaloniki, Greece
Jörg Waldvogel	Swiss Federal Institute of Technology, ETH, Switzerland

Further details on the Scottish Universities Summer School in Physics 62 can be found at http://http://www.astro.gla.ac.uk/users/martin/skye07/.

Executive Committee

Prof B A Steves	Glasgow Caledonian Univ.	*Director and Co-Editor*
Dr M Hendry	University of Glasgow	*Secretary and Co-Editor*
Prof A C Cameron	University of St Andrews	*Treasurer and Co-Editor*

International Advisory Committee

Prof B A Steves	Glasgow Caledonian Univ.	*Director and Co-Editor*
Prof A C Cameron	University of St Andrews, UK	
Prof Cl Froeschlé	Observatoire de la Cote d'Azur, Nice, France	
Dr M Hendry	University of Glasgow, UK	
Prof A Milani	Universitá di Pisa, Italy	

Preface

The detection and characterisation of extra-solar planets is one of the most active and exciting fields in astrophysics. Since the discovery in 1995 of the first exoplanet orbiting a main sequence star, over 300 planets have been detected and this number is expected to further increase dramatically in the next few years as a range of new ground-based planetary searches report their results. While the bulk of discoveries to date have been made via the Doppler 'wobble' technique, where the presence of a massive planet is detected from tiny periodic shifts in the wavelength of lines in the spectrum of its parent star, a growing number of planets have been found by other techniques - including the transit method, astrometric 'wobble' and gravitational microlensing. These techniques offer the tantalising possibility of the 'holy grail' of exoplanet searches: the detection of an Earth-mass planet in the habitable zone of a solar-type star. In the longer term lies the exciting prospect of carrying out spectroscopic study of exoplanetary atmospheres - observations which could reveal the presence of 'biomarkers', including water vapour, oxygen and carbon dioxide.

This textbook provides an invaluable reference volume for all students and researchers interested in studying the detection, formation, evolution and dynamics of planetary systems. Based on the lectures of the Scottish Universities Summer School in Physics entitled, 'The Extra Solar Planets: the detection, formation, evolution and dynamics of planetary systems', the book is written by internationally renowned scientists working at the forefront of the field. The textbook reviews current research and contains useful teaching on the latest tools and methods of analysis for investigating extra-solar planetary systems.

The rich harvest of new planets from the Doppler 'wobble' method has revealed the presence of a substantial population of 'hot Jupiters' - massive planets orbiting much closer to their parent star than even Mercury does in our solar system. While the discovery of these planets is to some extent a selection effect - since their mass and orbital radius produce a large Doppler wobble - their existence has nonetheless presented exciting challenges for theories of planetary formation based on planetesimal accretion. New theories on the evolution of planetary systems involving planetary migration and resonant capture are being developed to explain the existence of hot Jupiters, and the astrophysical mechanisms required to assemble gas giant planets close to their parent star are also an area of intense study.

The discovery of a wealth of new multi-planet systems is leading to new frontiers in the study of the dynamics of planetary systems. Mathematical tools involving periodicity, chaos and resonance are being used to explain the diversity of systems observed and to study their stability. Before the discovery of extra-solar planetary systems, dynamical study of planetary systems was focused on the solar system with its 8 planets and several large satellite systems typically exhibiting small eccentricities and near resonant orbits. New discoveries show planets of large mass, large eccentricity, orbiting in close orbits to their star. Strongly affected by gravitational interactions, these planets can only remain stable if trapped in mean motion resonances. Other planetary systems discovered with large period ratios exhibit apsidal locking due to significant non-resonant secular dynamics. The analysis of the transition from secular to resonant dynamics, apsidal corotation resonances, planetary migration due to tidal interaction, capture into corotation resonances, and evolution after capture are proving very fruitful in their explanation and understanding of the newly observed planetary systems.

The twelfth in the 'Cortina' series of Summer Schools began in the early 1970's on Dynamical Astronomy in Europe, SUSSP62 brought together around 55 scientists from over 20 countries. Held at Sabhal Mor Ostaig, Skye, Scotland, the two-week long summer school was a resounding success with the participants forming many new research links for the future, improving our understanding of our own solar system and its place in the diverse range of planetary systems discovered so far.

We would like to especially thank Dr Shoaib Afridi (University of Hail, Saudi Arabia), Dr Nikolaos Georgakarakos (ATEI of Western Macedonia, Greece) and Anoop Sivasankaran (Glasgow Caledonian University, UK) for their dedication and hard work which contributed so much to the success of the school and the final textbook. We are also indebted to SUSSP, SUPA, STFC, EPSRC and ESF for their valuable help and sponsorship. Further information on the Summer School can be found at: http://www.astro.gla.ac.uk /users/martin/skye07/. We hope you enjoy the text.

BA Steves

M Hendry

AC Cameron

March 2010

Contents

I Detection of Extra-Solar Planets: Methods and Observations   **1**

Detection of extra-solar planets in wide-field transit surveys   **3**
     1   Introduction 3
     2   Observations and data reduction 4
     3   Zero-point correction and data weights 5
     4   Removal of correlated systematic errors 6
     5   Characterising red noise 8
     6   Transit-search algorithms 9
     7   Estimation of system parameters 10
     8   Markov-chain Monte-Carlo modelling 12
     9   Candidate selection 14
     10   Conclusions 14

The theory of planet detection by gravitational microlensing: blips and dips from the gravitational bending of light   **17**
     1   Gravitational bending of light 17
     2   Gravitational microlensing 18
         2.1   Lens equation 18
         2.2   Image distortion and magnification 20
         2.3   Microlensing light curves 21
     3   Microlensing event rate 22
     4   Where to find planets 23
     5   Binary microlenses 25
     6   Planetary microlensing 28
         6.1   Images and caustics 28
         6.2   Excess magnification 28
         6.3   Light curves and planet detection 30

The practice of planet detection by gravitational microlensing: studying cool planets around low-mass stars   **35**
     1   Gravitational microlensing events 35
     2   Microlensing planet searches 36
     3   Detection efficiency and abundance limits 38
     4   The first planet detections 38
     5   Probing planet parameter space 41
     6   Planetary census 42
     7   Planets of Earth mass and below 44

Modelling spectroscopic and polarimetric signatures of exoplanets   **49**

1   Introduction 49
2   Describing and calculating planetary radiation 52
    2.1   Flux and polarization 52
    2.2   Planetary radiation 53
    2.3   Reflected starlight 56
    2.4   Radiative transfer parameters 58
    2.5   Radiative transfer algorithms 63
3   Flux and polarization spectra 64
    3.1   Observations of the Earth 64
    3.2   Simulations of exoplanet spectra 68
4   Summary 74

II   Formation and Evolution of Planetary Systems   **79**

Proto-planetary discs – current problems and directions   **81**

1   Introduction 81
2   Disc properties 83
    2.1   Disc masses and sizes 84
3   Effects of planet formation in discs 86
    3.1   Expected outcomes 87
    3.2   Imaging planet-formation 88
4   Summary 89

Debris discs and planetary environments   **91**

1   Introduction 91
2   Disc observations 93
    2.1   Examples 94
    2.2   Trends 96
3   Theoretical interpretation 97
4   Planetary signatures 98
5   Astrobiological implications 99
6   Summary and future 99

Dynamical evolution of planetary systems   **101**

1   Planet formation 101
    1.1   Constraints from observations of protoplanetary disks 102
    1.2   Classical planet formation via sequential accretion . . . 103
    1.3   Chaotic phase of planet formation 104
2   New wrinkles to planet formation theory 105
    2.1   Orbital Migration 105
    2.2   Eccentricity excitation 107
    2.3   Passing stars and wide binary companions 108

	2.4	Planet-planet interactions	108
	2.5	Additional proposed mechanisms	110
3		Multiple planet systems	110
	3.1	Classes of multiple planet systems	111
	3.2	Multiple planet systems: probes of planet formation	113
4		Future tests of planet formation models	115
	4.1	Radial velocity observations	115
	4.2	Transit searches	116
	4.3	Long-term	117

Late stages of solar system formation and implications for extra-solar systems   **123**

	1		Formation stages of planetary systems	123
	2		Planet migration	125
		2.1	Gas-driven migration	126
		2.2	Planetesimals-driven migration	131
	3		Probing the history of the solar system	132
		3.1	The Nice model	135
		3.2	Connection with the gas-rich era	140
	4		Implications for extra-solar systems	143
	5		Summary	145

III   Dynamics of Planetary Systems   **149**

A brief account of mutual planetary perturbations   **151**

1		Introduction	151
2		Review of Kepler two-body motion	152
	2.1	The orbit in general	152
	2.2	The elliptic orbit	153
	2.3	The orbit in time	154
	2.4	The orientation of the orbit in space	155
3		Perturbed elliptic motion	156
	3.1	The main features of the disturbing function	158
4		The canonical form of the equations of motion	159
	4.1	Lie series transformations	159
	4.2	A canonical formulation of an n-planet system	161
	4.3	Separation of the short-period terms from the long-period problem	162
	4.4	Solution of the long-period part	163
	4.5	Properties of the complete solution	166
	4.6	Small-integer commensurabilities	166

**Fundamentals of regularization in celestial mechanics and
linear perturbation theories**   **169**

1   Introduction 169
2   Planar Kepler motion 170
3   The Levi-Civita transformation 172
     3.1   First step: slow-motion movie 173
     3.2   Second step: conformal squaring 173
     3.3   Third step: fixing the energy 174
4   Spatial regularization with quaternions 175
     4.1   Basics 175
     4.2   The KS map in the language of quaternions 177
     4.3   The inverse map 178
     4.4   The regularization procedure with quaternions 178
5   The perturbed spatial Kepler problem 180
     5.1   Osculating elements 182
6   Conclusions 183

Mechanisms for the production of chaos in dynamical systems **185**

1   Introduction 185
2   The homoclinic tangle of hyperbolic saddle points 188
3   The Smale horseshoe 190
4   Chaotic dynamics in the homoclinic tangles 196
5   From chaos to diffusion in two dimensional systems 197
6   Diffusion in higher dimensional systems: the Arnold's model
     for diffusion 200
7   Arnold diffusion in a quasi integrable 4D system 201
8   An application to our planetary system 205

Extra-solar multiplanet systems   **211**

1   Introduction 211
2   Class I. Planets in close orbits 212
     2.1   Class Ia. Planets in resonant orbits (MMR) 215
     2.2   Class Ib. Low-eccentricity near-resonant pairs 217
3   Class II. Non-resonant planets with significant secular
     dynamics 219
4   Class III. Weakly-interacting planet pairs 221
5   Equations of motion for N planets 222
     5.1   Astrocentric equations of motion 223
6   Reduction of the Hamiltonian equations 223
     6.1   Jacobi's canonical coordinates 224
     6.2   Poincaré's relative canonical coordinates 227
7   Action-angle variables: Delaunay elements 228
8   Keplerian elements 231
     8.1   Application to Jacobi's and Poincaré's canonical
         variables 232

9   Kepler's third law 233
10   Conservation of angular momentum 234

The stability of terrestrial planets in planetary systems   **241**
1   Introduction 241
2   The dynamical models 241
3   Numerical methods and analysis of the results 244
4   Regions of motion of terrestrial planets in habitable zones
      of EPS 246
      4.1   Terrestrial planets in the IHZ 246
      4.2   Terrestrial planets in the THZ 247
      4.3   Terrestrial planets in double stars 249
5   Summary 257

Did the two Earth poles move widely 13,000 years ago? An
astrodynamical study of the Earth's rotation   **259**
1   Introduction 259
      1.1   Preliminary notice (numeration system) 260
2   Why this study? 261
      2.1   The geographical extension of the last ice age 261
      2.2   The concept of a stable equilibrium position of
             the poles 262
3   The matrix M_c before the melting of the last ice age 271
      3.1   The effect of isostasy 271
      3.2   Estimation of the Earth inertia matrix M_{C1} during the
             last ice age 277
4   Are such fast and large moves of the poles possible? 282
5   The age of natural disasters 282
6   Conclusion 283

List of Participants   **285**

Index   **287**

Part I

Detection of Extra-Solar Planets: Methods and Observations

Detection of extra-solar planets in wide-field transit surveys

Andrew Collier Cameron

SUPA, School of Physics and Astronomy, University of St Andrews, Scotland

In this lecture I describe the observational and transit-detection strategies adopted by current wide-field surveys aimed at discovering bright transiting gas-giant planets using arrays of small cameras with apertures of order 11 cm. The advantage of using such small instruments is that their wide sky coverage permits discovery of relatively bright systems amenable to detailed follow-up studies.

I present algorithms commonly used to identify and remove sources of systematic error from the data, and to detect the presence of periodic transit signals. I shall then discuss methods for identifying and eliminating astrophysical false positives using the photometric discovery data combined with publicly-available photometric and astrometric catalogue information.

1 Introduction

Among more than 300 extra-solar planets that have been discovered since 1995, the 35 that transit their parent stars offer the greatest physical insights into their structure and evolution. The geometry of a transit, combined with radial-velocity measurements of the star's reflex orbit, yields direct measurements of the stellar density and the planetary surface gravity (Southworth et al 2007). Other parameters of interest, including the planetary density, can be determined once the stellar mass is known. The planetary density is an important diagnostic of the mass of a planet's rock/ice core (Fortney et al 2007, Seager et al 2008), and hence of its formation history.

Some of the most spectacular advances in our understanding of the thermal and chemical make-up of hot-Jupiter atmospheres are being achieved through optical (Pont et al 2008) and infrared photometry (Knutson et al 2008) and spectroscopy (Tinetti et al 2008) of both primary and secondary transits with HST and Spitzer. The wavelength dependence of the planet's silhouette during primary transit, and the spectrum of the flux deficit that occurs as the planet passes behind the star, depend on the molecular composition and vertical temperature profile of the atmosphere above a dusty cloud deck. Since both methods rely on subtraction of large numbers to identify a faint difference signal, only the brightest transiting planets are amenable to this sort of study.

2   Observations and data reduction

At the time of writing, four major wide-field surveys for transiting extra-solar planets are in progress. The Transatlantic Exoplanet Survey, (TrES: Dunham et al 2004; Alonso et al 2004) is the longest-running of these projects, and comprises a network of three wide-field survey cameras of 10-cm aperture: the STARE camera on Tenerife, the PSST project at Lowell Observatory, Arizona, and the SLEUTH project on Palomar Mountain in California. The HATnet project is a network of fully-automated telescopes whose commercial 200mm, f/1.8 lenses give 8x8-degree fields of view, located at the F L Whipple Observatory, Arizona and on Mauna Kea, Hawaii (Bakos et al 2002). The XO project (McCullough et al 2005) uses two cameras utilising the same 200mm, f/1.8 lenses as the HATnet project, atop Haleakala on the island of Maui, Hawaii. The WASP project (Pollacco 2006) operates two arrays of 8 cameras: one atop La Palma, Canary Islands, and one at the Sutherland observing station of the South African Astronomical Observatory. The project utilises the same lenses as HATnet and XO, backed by Peltier-cooled CCD arrays of 2048^2 pixels.

Observing strategies for these projects vary. HATnet and TrES have the advantage of multiple longitude coverage, and adopt a strategy in which each camera observes one field continuously. XO and WASP adopt sky-tiling strategies, trading cadence for more complete coverage of the sky. Both installations return to the same field once every few minutes; each field is observed for up to 8 hours per night throughout an observing season of typical duration 4 months. The duration of the observing season is determined by the requirement that each field should be observable for at least 4 hours per night.

The large volumes of data generated by these robotic installations require continuous unattended pipeline data reduction. A common feature of the commercial camera lenses used by these projects is that they suffer from significant vignetting. Great care must be taken to secure good-quality flat fields using the twilight sky at dawn and dusk as often as conditions permit. Standard procedures for CCD image processing (bias and dark removal, flat-fielding to correct for vignetting and pixel-to-pixel variations) are applied. Light curves are extracted from the processed images using either simple aperture photometry at the locations of known catalogued objects (HAT, XO, WASP) or differential image analysis (TrES). The discussion that follows is specific to the WASP project, but similar methods are employed by all the current surveys.

3 Zero-point correction and data weights

The primary and secondary extinction coefficients and the colour equation and zero point for each night's data are determined from the correlations between instrumental magnitude, airmass and stellar magnitudes and colours from the TYCHO-2 catalogue (Hog et al 2000).

Each night's data is calibrated independently. Once a season's data has been processed, it is important to correct for time-dependent drifts in the photometric zero point using the entire season's data. We start with a two-dimensional array m_{ij} of processed stellar magnitudes from the pipeline. The first index i denotes a single CCD frame within the entire season's data. The second index labels an individual star. We compute the mean magnitude of each star:

$$\hat{m}_j = \frac{\sum_i m_{ij} w_{ij}}{\sum_i w_{ij}} \tag{1}$$

where the weights w_{ij} incorporate both the formal variance σ_{ij}^2 calculated by the pipeline from the stellar and sky-background fluxes, and an additional systematic variance component $\sigma_{t(i)}^2$ introduced in individual frames by passing wisps of cloud, stratospheric dust events and other transient phenomena which degrade the extinction correction:

$$w_{ij} = \frac{1}{\sigma_{ij}^2 + \sigma_{t(i)}^2}. \tag{2}$$

Data points from frames of dubious quality are thus down-weighted. The weight is set to zero for any data point flagged by the pipeline as either missing or bad.

The zero-point correction for each frame i follows:

$$\hat{z}_i = \frac{\sum_j (m_{ij} - \hat{m}_j) u_{ij}}{\sum_j u_{ij}} \tag{3}$$

In this case the weights are defined as

$$`u_{ij} = \frac{1}{\sigma_{ij}^2 + \sigma_{s(j)}^2}, \tag{4}$$

where $\sigma_{s(j)}^2$ is an additional variance caused by intrinsic stellar variability. This down-weights variable stars in the calculation of the zero-point offset for each frame.

Initially we set $\sigma_{t(i)}^2 = \sigma_{s(j)}^2 = 0$, and compute the average magnitude $\hat{m}_j$ for every star j and the zero-point offset $\hat{z}_i$ for every frame.

The additional variance of $\sigma_{s(j)}^2$ for the intrinsic variability of a given star j is determined from the data themselves using a maximum-likelihood (ML) approach (Collier Cameron et al 2006); the additional variance $\sigma_{t(i)}^2$ generated by transparency fluctuations within a given image is computed in an analogous manner.

At this stage we refine the mean magnitude per star, the zero point offsets and the additional variances for stellar variability and patchy cloud, by iterating Eqs. (1), (3) and the ML estimates of $\sigma^2_{s(j)}$ and $\sigma^2_{t(i)}$ to convergence. The corrected differential magnitude of each star is then given by

$$x_{ij} = m_{ij} - \hat{m}_j - \hat{z}_i. \tag{5}$$

4   Removal of correlated systematic errors

Several recognised sources of systematic error affect ultra-wide-field photometry obtained with commercial camera lenses. For example, the WASP bandpass spans the visible spectrum, thereby introducing significant colour-dependent terms into the extinction correction. The pipeline corrects for secondary extinction using TYCHO-2 $B - V$ colours for the brighter stars, but uncertainties in the colours of the TYCHO-2 stars and the lack of colour information for the fainter stars means that some systematic errors remain. Bright moonlight and stratospheric dust reduce the contrast between faint stars and the sky background, altering the rejection threshold for faint sources in the sky-background annulus and biassing the photometry for faint stars. The WASP camera lenses are vignetted across the entire field of view, and the camera array is not autoguided, so polar-axis misalignment causes stellar images to drift by a few tens of pixels across the CCD each night. Systematic errors can therefore arise if the vignetting correction is imperfect. Temperature changes during the night affect the camera focus, changing the shape of the point-spread function across the field and biassing the photometry for fainter stars. These systematic errors, and no doubt others as-yet unidentified, have a serious impact on the detection threshold for transits.

We use the SYSREM algorithm of Tamuz et al (2005) to identify and remove remaining patterns of correlated noise in the data. The SYSREM algorithm produces a corrected magnitude $\tilde{x}_{i,j}$ for star j at time i, given by

$$\tilde{x}_{i,j} = x_{i,j} - \sum_{k=1}^{M} {}^{(k)}c_j \, {}^{(k)}a_i, \tag{6}$$

where M represents the number of basis functions (each representing a distinct pattern of systematic error) removed. Each basis function a_i has the dimensions of a light curve, whose contribution to the light curve of star j is scaled by the coefficient c_j. The basis functions and coefficients are computed by minimising the badness-of-fit statistic

$$\chi^2 = \sum_{ij} \frac{m_{ij} - c_j a_i}{\sigma^2_{ij}} \tag{7}$$

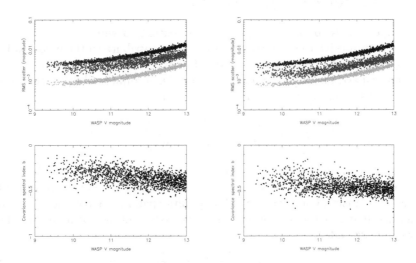

Figure 1. *Upper panel: RMS scatter versus magnitude before (left) and after (right) removal of the four strongest correlated error components using* SysRem. *The upper curve shows the RMS scatter in the decorrelated light-curves of the non-variable stars in the ensemble. The middle curve shows the scatter in the same light-curves after boxcar smoothing over all complete 2.5-hour intervals within each night. The lower curve shows the RMS scatter of the individual data points divided by the square root of the average number of points (typically 22) in a 2.5-hour interval. Lower panel: Covariance spectral index β as a function of V magnitude for the decorrelated data.*

with respect to c_j and a_i alternately. The process is found to iterate rapidly to convergence. Successive basis functions are computed by subtracting the array $c_j a_i$ from the data and repeating the process. The reader is referred to Tamuz et al (2005) for details of the algorithm, and to Collier Cameron et al (2006) for additional information concerning the implementation used for the WASP project. In the latter, additional variance components are computed per star and per image, to down-weight the light curves of variable stars and contributions from images taken in poor observing conditions.

Another common decorrelation method is the Trend-Filtering Algorithm (TFA: Kovács et al. 2005). This algorithm also searches for patterns of systematic variability common to large numbers of stars in a given field, constructing a set of basis functions from a set of template light curves distributed randomly with respect to position on the image.

5 Characterising red noise

It is important to be able to quantify the extent to which correlated noise sources that survive the decorrelation process contribute to the error budget. Pont et al. (2006) discussed methods of characterising the structure of the covariance matrix for a given stellar light-curve. A Jupiter-like planet orbiting a solar-like star has a transit duration of order

$$t_{dur} \simeq 2.5 \left(\frac{M_*}{M_\odot} \right)^{2/3} \left(\frac{P}{2.5 \text{ days}} \right)^{1/3} \text{ hours.} \tag{8}$$

The simplest method is to carry out boxcar smoothing of each night's data, with a smoothing length comparable to the typical 2.5-hour duration of a planetary transit. For every set of L points spanning a complete 2.5-hour interval starting at the kth observation, we construct an optimally-weighted average magnitude

$$\hat{m}_k = \frac{\sum_{i=k}^{k+L-1} m_i w_i}{\sum_{i=k}^{k+L-1} w_i}, \tag{9}$$

with bad observations down-weighted as above using $w_i = 1/(\sigma_i^2 + \sigma_{t(i)}^2)$.

For uncorrelated noise, we expect $\sigma_{\text{binned}} = \sigma_{\text{unbinned}}/\sqrt{L}$, where L is the average number of observations made in a 2.5-hour interval. The RMS scatter σ_{binned} in the smoothed light-curve of $\hat{m}_k$ values is then compared to the RMS scatter σ_{unbinned} of the individual data points as shown in Figure 1. The correlated noise amplitude of the binned data is reduced from 0.0025 to 0.0015 magnitude for the brightest stars, which despite SYSREM's best efforts is still significantly greater than the 0.0008 magnitude that would be achieved if the noise were uncorrelated.

The covariance structure of the correlated noise is quantified by the power-law dependence of the RMS scatter on the number of observations used in the boxcar smoothing:

$$\sigma_{\text{binned}} = \sigma_{\text{unbinned}} L^\beta. \tag{10}$$

For completely uncorrelated noise we expect $\beta = -1/2$, while for completely correlated (red) noise we expect the RMS scatter to be independent of the number of data points, giving $b = 0$. We measure β for each star using the incomplete smoothing intervals at the start and end of the night. We create a set of binned magnitudes obtained for $L = 1, 2, 3, \ldots$ consecutive observations. We compute the RMS scatter in the binned magnitudes for each value of N as a function of L and fit a power-law to determine β. In Figure 1 we plot β as a function of V magnitude, excluding intrinsic variable stars having $\sqrt{\sigma_{s(j)}^2} > 0.005$ magnitude. While some effects of correlated noise remain for the brightest stars, stars fainter than $V = 11.0$ have covariance spectral indices close to the value $\beta = -0.5$ expected for white noise.

6   Transit-search algorithms

The wide-field transit-search projects currently in operation all use variants of the Box-Least-Squares (BLS) algorithm of Kovács et al (2002) to search for objects displaying the periodic dimmings produced by planetary transits. Searches are conducted on a grid of orbital frequencies, using frequency increments such that the accumulated phase difference between successive frequencies over the full duration of the dataset corresponds to some fraction of the expected width of a transit at the longest period searched. A set of transit epochs is defined at each frequency, at phase intervals equal to a fraction of the expected transit width at that frequency. The expected transit duration is computed for each trial period using Kepler's third law assuming a stellar mass of 0.9 $M_\odot$. Since the majority of the main-sequence stars in the magnitude range of interest have masses between 0.7 and 1.3 $M_\odot$ and the transit duration at a given period scales as $(M_*/M_\odot)^{2/3}$, the predicted transit duration is unlikely to be in error by more than 20 to 30 percent even at the extremes of the mass range.

At each trial period and epoch, the transit depth and goodness-of-fit statistic are calculated. In the WASP project we use a variant of the BLS method developed by Aigrain & Irwin (2004) and by Burke et al (2007), in which the goodness-of-fit criterion has the dimensions of the χ^2 statistic.

The light-curve of a given star comprises a set of observations $\tilde{x}_i$ with associated formal variance estimates σ_i^2 and additional, independent variances $\sigma_{t(i)}^2$ to account for transient spatial irregularities in atmospheric extinction. We define inverse-variance weights

$$w_i = \frac{1}{\sigma_i^2 + \sigma_{t(i)}^2 + \sigma_{s(j)}^2}, \tag{11}$$

and subtract the optimal average value

$$\hat{x} = \frac{\sum_i \tilde{x}_i w_i}{\sum_i w_i} \tag{12}$$

to obtain $x_i = \tilde{x}_i - \hat{x}$. We also define

$$t = \sum_i w_i, \quad \chi_0^2 = \sum_i x_i^2 w_i, \tag{13}$$

summing over the full dataset. Note that the weights defined here include the independent variance component $\sigma_{s(j)}^2$. This has the effect of lowering the significance of high-amplitude variable stars, but has little effect on low-amplitude variables such as planetary transit candidates.

In the BLS method, the transit is modelled by a periodic box function whose period, phase and duration determine the subset ℓ of "low" points observed during transits. We partition the data into a contiguous block of out-of-transit points for which $w/2P < \phi < 1 - w/2P$, where w is the transit duration and P is the orbital period, and the complement of this subset comprising the in-transit points. In many implementations the partitioning of the data can be the slowest part of the BLS procedure. Substantial speed gains can be achieved by judicious ordering and binning of the data.

Using notation similar to that of Kovács et al (2002), we define

$$s = \sum_{i \in \ell} x_i w_i, \quad r = \sum_{i \in \ell} w_i, \quad q = \sum_{i \in \ell} x_i^2 w_i. \tag{14}$$

The mean light levels inside (L) and outside (H) transit are given by

$$L = \frac{s}{r}, H = \frac{-s}{t-r} \tag{15}$$

with associated variances

$$\mathrm{Var}(L) = \frac{1}{r}, \quad \mathrm{Var}(H) = \frac{1}{t-r}. \tag{16}$$

The fitted transit depth and its associated variance are

$$\delta = L - H = \frac{st}{r(t-r)}, \quad \mathrm{Var}(\delta) = \frac{t}{r(t-r)}, \tag{17}$$

so the signal-to-noise ratio of the transit depth is

$$\mathrm{S/N} = s\sqrt{\frac{t}{r(t-r)}}. \tag{18}$$

The improved fit to the data is given by $\chi^2 = \chi_0^2 - \Delta\chi^2$, where the improvement in the fit when compared with that of a constant light curve is

$$\Delta\chi^2 = \frac{s^2 t}{r(t-r)}. \tag{19}$$

Note also that $\Delta\chi^2 = s\delta = (S/N)^2$. The goodness of fit to the portions of the light-curve outside transit, where the light level should be constant, is

$$\chi_h^2 = \chi_0^2 - \frac{s^2}{(t-r)} - q. \tag{20}$$

The best-fitting model at each frequency is selected, and the corresponding transit depths, $\Delta\chi^2$ and χ_h^2, are stored for each star.

7   Estimation of system parameters

Following an initial, coarse application of the BLS algorithm, we filter the candidates by rejecting obviously variable stars for which the post-fit $\chi^2 >$

$3.5N$, where N is the number of observations. We reject stars for which the best solution has fewer than two transits.

We select candidates according to the "signal-to-red noise" ratio S_{red} of the best-fit transit depth to the RMS scatter binned on the expected transit duration:

$$S_{red} = \frac{\delta\sqrt{N_t}}{\sigma L^\beta}. \tag{21}$$

As in Section 5 above, L is the average number of data points spanning a single transit, β is the power-law index that quantifies the covariance structure of the correlated noise, N_t is the number of transits observed, δ is the transit depth and σ is the weighted RMS scatter of the unbinned data.

The reduced sample is again subjected to a BLS search, this time utilising a finer grid spacing. For each star in the sample we identify the five most significant peaks that correspond to transit-like dimmings, and apply a Newton-Raphson method based on an analytic transit-profile model to determine the best-fitting width, depth, epoch and period of the putative transit signal.

At this stage it is possible to estimate the radii R_* of the host star and R_p of the transiting object from the light curve alone.

The physical parameters R_*/a, R_p/R_*, b and P, together with limb-darkening coefficients appropriate to the stellar effective temperature, determine completely the form and duration of the transit profile. We parametrise the form of the transit profile in terms of four observable parameters that are more directly determined by the observations: the fractional flux deficit ΔF at mid-transit, the total transit duration t_T from first to last contact, the impact parameter b of the planet's trajectory to the centre of the star in units of the stellar radius, and the orbital period P.

Using the approximation that $R_p + R_* << a$ we determine the stellar radius from the transit depth, duration and impact parameter via the relation:

$$\frac{R_*}{a} = \frac{t_T}{P}\frac{\pi}{(1+\sqrt{\Delta F})^2 - b^2}. \tag{22}$$

For transits where the companion is fully silhouetted against the primary during the middle part of the transit we neglect limb darkening and define

$$\frac{R_p}{R_*} = \sqrt{\Delta F}. \tag{23}$$

We use Kepler's third law to determine the stellar density in solar units:

$$\frac{\rho}{\rho_\odot} = 0.0134063\left(\frac{a}{R_*}\right)\left(\frac{P}{1\text{day}}\right)^{-2}. \tag{24}$$

In order to close the system and determine the stellar radius in solar units, we need either an additional equation or an independent determination of the stellar mass. We use the $J - H$ colour index taken from the 2MASS Point Source Catalogue (Cutri et al 2003) to estimate the stellar effective temperature and hence the corresponding main-sequence stellar mass (Collier Cameron et al 2007). The stellar radius then follows from the stellar mass and density.

8   Markov-chain Monte-Carlo modelling

Markov-chain Monte-Carlo (MCMC) methods are rapidly gaining popularity for solving multivariate parameter-fitting problems in astronomy and many other branches of science. They not only optimise the fit of a model to data, but they yield the joint posterior probability distribution of the fitted parameters. MCMC methods are particularly well-suited to the problem of deriving the physical parameters of star-planet systems by optimising model fits to the light curves of transiting exoplanets (Holman et al 2006).

Here we assume the planet's orbit to be circular. We characterise the system using the six parameters $\{T_0, P, \Delta F, t_T, b, M_*\}$. Here these six quantities constitute the "proposal parameters", which are allowed to perform a random walk through parameter space, generating a cloud of points that map out the joint posterior probability distribution.

At each step in the MCMC procedure, each proposal parameter is perturbed from its previous value by a small random amount:

$$
\begin{aligned}
T_{0,i} &= T_{0,i-1} + \sigma_{T_0} G(0,1) f \\
P_i &= P_{i-1} + \sigma_P G(0,1) f \\
\Delta F_i &= \Delta F_{i-1} + \sigma_{\Delta F} G(0,1) f \\
t_{T,i} &= t_{T,i-1} + \sigma_{t_T} G(0,1) f \\
b_i &= T_{0,i-1} + \sigma_b G(0,1) f \\
M_{*,i} &= M_{*,i-1} + \sigma_M G(0,1) f
\end{aligned}
$$

where $G(0,1)$ is a random Gaussian deviate with mean zero and unit standard deviation. The scale factor f is an adaptive step-size controller of order unity (Collier Cameron et al 2007).

The first four parameters (T_0, P, ΔF and t_T) and their associated one-sigma uncertainties are taken directly from the initial fit to the light curve. The impact parameter is given an initial value $b_0 = 0.5$ and a one-sigma uncertainty $\sigma_b = 0.05$. The stellar mass M_* is initially set to the value M_0 derived from the $J - H$ colour using the calibration described by (Collier Cameron et al 2007), and assigned an arbitrary but plausible one-sigma uncertainty $\sigma_M = 0.1 M_0$.

Once the physical parameters R_*, R_p, a and $\cos i$ have been derived from the proposal parameters ΔF, t_T, b and P using the relationships above, the projected separation of centres (in units of the primary radius) at any time t_j

of observation is

$$z(t_j) = \frac{\sin^2 \phi_j + (bR_*/a)^2 \cos^2 \phi_j}{R_*/a}. \tag{25}$$

The orbital phase angle at time t_j is $\phi_j = 2\pi(t_j - T_0)/P$.

We compute the flux deficit at each phase of observation using an analytic model (Mandel & Agol 2002) for small planets with an appropriate 4-coefficient limb-darkening model (Claret 2000). After converting these flux deficits to model magnitudes μ_j relative to the flux received from the system outside transit, we compute the zero-point offset from the observed magnitudes m_j:

$$\Delta m = \frac{\sum_j (m_j - \mu_j) w_j}{\sum_j w_j}, \tag{26}$$

where the observational errors σ_j define the inverse-variance weights $w_j = 1/\sigma_j^2$. We thus obtain the fitting statistic for the set of model parameters pertaining to the ith step of the Markov chain:

$$\chi_i^2(T_0, P, \Delta F, t_T, b, M_*) = \sum_j \frac{(m_j - \mu_j - \Delta m)^2}{\sigma_j^2}. \tag{27}$$

The likelihood of obtaining the observed data D given the model defined by a particular set of proposal parameters is

$$\mathcal{P}(D|T_0, P, \Delta F, t_T, b, M_*) \propto \exp(-\chi^2/2), \tag{28}$$

but the full posterior probability distribution for the data and the model depends on the prior probability distribution for each of the model parameters. We are only interested in solutions for which the companion is a planet-sized object yielding a transit of observable depth, so we restrict the impact parameter to the range $0 < b < 1.0$, rejecting proposal steps that fall outside this range.

For most of the remaining parameters the uniform prior implied by the random-walk nature of MCMC is valid. The stellar mass and radius are already determined from the $J - H$ colour, under the assumption that the star is single and on the main sequence. Under this prior assumption we expect the stellar mass to lie somewhere within an approximately gaussian distribution with mean M_0 and standard deviation $\sigma_M = M_0/10$ (i.e. the same arbitrary but plausible value used to determine the average jump size in M_*). We use a power-law approximation to the main-sequence mass-radius relation to define a prior probability distribution for R_* with mean $R_0 - M_0^{0.8}$ (Tingley & Sackett 2005) and hence standard deviation $\sigma_R = 0.8(R_0/M_0)\sigma_M$.

This gives a joint prior probability distribution for the values of the proposal parameter M_* and the derived physical parameter R_* of the form

$$\mathcal{P}(M_{*,i}, R_{*,i}) = \exp\left(-\frac{(M_{*,i} - M_0)^2}{2\sigma_M^2} - \frac{(R_{*,i} - R_0)^2}{2\sigma_R^2}\right). \tag{29}$$

Since the posterior probability distribution is $\mathcal{P}(M_{*,i}, R_{*,i})\mathcal{P}(D|T_0, P, \Delta F, t_T, b, M_*)$, we impose the prior on $M_{*,i}$ and $R_{*,i}$ by replacing χ_i^2 with the logarithm of the posterior probability distribution

$$Q_i \equiv \chi_i^2 + \frac{(M_{*,i} - M_0)^2}{\sigma_M^2} + \frac{(R_{*,i} - R_0)^2}{\sigma_R^2} \tag{30}$$

as the statistic on which acceptance of a set of proposal parameters is decided.

For every new proposal set generated, the decision as to whether or not to accept the set is made via the Metropolis-Hastings rule: if $Q_i < Q_{i-1}$ the new set is accepted; if on the other hand $Q_i > Q_{i-1}$, the new set is accepted with probability $\exp(-\Delta Q/2)$, where $\Delta Q \equiv Q_i - Q_{i-1}$. The algorithm first converges to, then explores the parameter space around a constrained optimum solution that represents a compromise between fitting the light curve and reconciling the resulting stellar dimensions with prior expectations derived from the $J - H$ colour.

9   Candidate selection

The MCMC method has the significant advantage that it allows us to select candidates for which there is a significant probability that the transiting companion has a planet-like radius. The fraction of trials in the Markov chain for which the planet radius $R_p < 1.5 R_{\mathrm{Jup}}$ is a direct estimate of this probability.

The value of the main-sequence prior $\mathcal{P}(M_{*,i}, R_{*,i})$ at the global minimum of Q constitutes a useful measure of how far the stellar parameters must be displaced from the main sequence values appropriate to the star's colour in order to fit the transit light curve. Astrophysical false positives such as stellar binaries and bright stars blended with background binaries tend to show transit durations that are inconsistent with the main-sequence stellar mass and radius implied by the $J - H$ colour. Promising planet candidates therefore tend to have low values of the statistic

$$S = \frac{(M_{*,i} - M_0)^2}{\sigma_M^2} + \frac{(R_{*,i} - R_0)^2}{\sigma_R^2}. \tag{31}$$

We are therefore able to winnow the sample, picking out only those objects that have a significant probability of having planet-sized companions transiting stars whose radii and masses appear to be consistent with their $J - H$ colours.

10   Conclusions

The methodology outlined in this chapter has been applied in practice by the WASP project to select target lists for radial-velocity follow-up studies using the SOPHIE spectrograph on the 1.93-m telescope at the Observatoire de Haute-Provence, and with the CORALIE spectrograph on the 1.2-m Swiss Euler telescope at La Silla. The candidate selection procedures derived here have yielded 20 confirmed planet discoveries at the time of writing, with a success rate of approximately one planet per 5 or 6 candidates examined. The main types of astrophysical false positive that elude the winnowing procedures described here are hierarchical triples (in which a bright main-sequence star is accompanied by a physically-bound pair of low-mass stars in a short-period eclipsing binary) and stellar binaries in which the secondary is a cool, low-mass main-sequence star with a radius similar to that of a large gas-giant planet. The discovery rate of transiting planets is expected to accelerate further as more efficient means are found of eliminating systematic errors, conducting transit searches, and eliminating astrophysical impostors from target lists.

References

Aigrain, S. & Irwin, M. (2004), 'Practical planet prospecting', MNRAS **350**, 331–345.

Alonso, R. et al. (2004), 'TrES-1: The transiting planet of a bright K0 V Star', ApJ **613**, L153–L156.

Bakos, G. Á. et al. (2002), 'System description and first Light curves of the Hungarian automated telescope, an autonomous observatory for variability search', PASP **114**, 974–987.

Burke, C. et al. (2007), 'XO-2b: Transiting hot Jupiter in a metal-rich common proper motion binary', ApJ **671**, 2115–2128.

Claret, A. (2000), 'A new non-linear limb-darkening law for LTE stellar atmosphere models. Calculations for $-5.0 \leq \log[M/H] \leq +1$, $2000K \leq T_{eff} \leq 50000$ K at several surface gravities', A&A **363**, 1081–1190.

Collier Cameron, A. et al. (2006), 'A fast hybrid algorithm for exoplanetary transit searches', MNRAS **373**, 799–810.

Collier Cameron, A. et al. (2007), 'Efficient identification of exoplanetary transit candidates from SuperWASP light curves', MNRAS **380**, 1230–1244.

Cutri, R. et al. (2003), *2MASS All Sky Catalog of point sources*, The IRSA 2MASS all-sky point source catalog, NASA/IPAC Infrared Science Archive. http://irsa.ipac.caltech.edu/applications/Gator/.

Dunham, E., et al. (2004), 'PSST: The Planet Search Survey Telescope', PASP **116**, 1072–1080.

Fortney, J., Marley, M. & Barnes, J. (2007), 'Planetary radii across five orders of magnitude in mass and stellar insolation: Application to transits', ApJ **659**, 1661–1672.

Hog, E. et al. (2000), 'The Tycho-2 Catalogue (Hog+ 2000)', VizieR Online Data Catalog **1259**.

Holman, M. et al. (2006), 'The Transit Light Curve Project. I. Four consecutive transits of the exoplanet XO-1b', ApJ **652**, 1715–1723.

Knutson, H. et al. (2007), 'A map of the day-night contrast of the extra-solar planet HD 189733b', Nature **447**, 183–186.

Kovács, G., Bakos, G. & Noyes, R. (2005), 'A trend filtering algorithm for wide-field variability surveys', MNRAS **356**, 557–567.

Kovács, G., Zucker, S. & Mazeh, T. (2002), 'A box-fitting algorithm in the search for periodic transits', A&A **391**, 369–377.

Mandel, K. & Agol, E. (2002), 'Analytic light curves for planetary transit Searches', ApJ **580**, L171–L175.

McCullough, P. et al. (2005), 'The XO Project: Searching for transiting extra-solar planet candidates', PASP **117**, 783–795.

Pollacco, D. o. (2006), 'The WASP project and the SuperWASP cameras', PASP **118**, 1407–1418.

Pont, F., Zucker, S. & Queloz, D. (2006), 'The effect of red noise on planetary transit detection', MNRAS **373**, 231–242.

Pont, F. et al. (2008), 'Detection of atmospheric haze on an extra-solar planet: the 0.55-1.05 μm transmission spectrum of HD 189733b with the Hubble Space Telescope', MNRAS **385**, 109–118.

Protopapas, P., Jimenez, R. & Alcock, C. (2005), 'Fast identification of transits from light-curves', MNRAS **362**, 460–468.

Schwarzenberg-Czerny, A. & Beaulieu, J. (2006), 'Efficient analysis in planet transit surveys', MNRAS **365**, 165–170.

Seager, S. & Mallén-Ornelas, G. (2003), 'A unique solution of planet and star parameters from an extra-solar planet transit light curve', ApJ **585**, 1038–1055.

Seager, S. et al. (2007), 'Mass-radius relationships for solid exoplanets', ApJ **669**, 1279–1297.

Southworth, J., Wheatley, P. & Sams, G. (2007), 'A method for the direct determination of the surface gravities of transiting extra-solar planets', MNRAS **379**, L11–L15.

Tamuz, O., Mazeh, T. & Zucker, S. (2005), 'Correcting systematic effects in a large set of photometric light curves', MNRAS **356**, 1466–1470.

Tinetti, G. et al. (2007), 'Water vapour in the atmosphere of a transiting extra-solar planet', Nature **448**, 169–171.

Tingley, B. & Sackett, P. (2005), 'A photometric diagnostic to aid in the identification of transiting extra-solar planets', ApJ **627**, 1011–1018.

The theory of planet detection by gravitational microlensing: blips and dips from the gravitational bending of light

Martin Dominik

SUPA, University of St Andrews, School of Physics & Astronomy, North Haugh, St Andrews, KY16 9SS, United Kingdom

Royal Society University Research Fellow

Introduction

In 1912, Albert Einstein realized that a distant star can exhibit a transient brightening caused by the bending of light due to the gravitational field of an intervening foreground star. However, because of the rarity of sufficiently close angular alignments, it required several decades of advances in technology until the observation of such gravitational microlensing events turned into reality. Planets around the lens star can reveal their existence with finite probability by causing blips or dips to microlensing light curves that are otherwise symmetric around a peak. Such planetary signals last between hours and weeks depending on the mass of the planet as well as the angular size of the observed source star and its proper motion with respect to the foreground 'lens' star. Due to a resonance effect between the deflections caused by the lens star and its planets, the latter are more easily detected if their angular separation falls into the 'lensing zone'.

1   Gravitational bending of light

From the equivalence principle, Einstein deduced that gravity should bend light, where the light ray follows a hyperbola. For an impact parameter ξ, a body of mass M with gravitational radius

$$R_S = \frac{2GM}{c^2},\qquad(1)$$

where G denotes the universal gravitational constant, and c is the vacuum speed of light, was thought to cause a total deflection by

$$\alpha_{NR} = \frac{R_S}{\xi}.\qquad(2)$$

He brought forward the suggestion to measure this bending near the solar limb, where $\alpha_{NR} = 0.83''$, during a solar eclipse (Einstein 1911). However, with the full theory of General Relativity (Einstein 1915), he found that the space-time curvature accounts for the deflection being twice as strong

$$\alpha = \frac{2\,R_S}{\xi},\qquad(3)$$

so that at the Solar limb $\alpha = 1.7''$.

Arthur Eddington, who was going to lead an expedition aiming at observing the Solar Eclipse of May 29th, 1919, stated three possible outcomes (Eddington 1919): 'The present eclipse expeditions may for the first time demonstrate the weight of light; or they may confirm Einstein's weird theory of non-Euclidean space; or they may lead to a result of yet more far-reaching consequences – no deflection.' However, Eddington's measurements confirmed the value predicted by Einstein's theory. In a discussion at the Royal Astronomical Society in the same year, Eddington praised it with the words 'The generalized relativity theory is a most profound theory of Nature, embracing almost all the phenomena of physics.' (Eddington et al. 1919).

2   Gravitational microlensing

2.1   Lens equation

Gravitational microlensing is understood as the bending of the light received from an observed distant (source) star due to the gravitational field of an intervening foreground (lens) star. The underlying theory is remarkably simple with all the physics absorbed into the deflection angle and the rest being geometry. The spherical symmetry of the space-time metric implies that angular momentum is conserved, so that the light rays remain in the plane defined by source, observer, and lens star.

The left panel of Figure 1 provides a side view. Measured with respect to a reference axis through observer and lens star, the light ray is separated by η at the source distance D_S and by ξ at the lens distance D_L.

Assuming that all angles are tiny, one finds the gravitational lens equation

$$\eta = \frac{D_S}{D_L}\xi - (D_S - D_L)\,\alpha(\xi).\qquad(4)$$

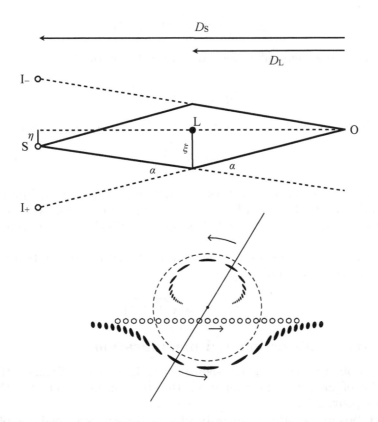

Figure 1. *Top side view illustrating the bending of light due to a massive object ('gravitational lens' L), which leads to an observer O receiving it from the direction of two apparent image positions* I_+ *and* I_- *on either side of the source star S. With respect to a reference axis through observer and lens,* η *and* ξ *denote the separation of the light ray at the source distance* D_S *or the lens distance* D_L*, respectively, while* α *is the bending angle, given by Eq. (3). Bottom: Observer's view of the two images of a circular source for different phases of a microlensing event, revealing the magnification and distortion. The dashed line indicates the Einstein circle (of angular radius* θ_E*) that belongs to the lens, which is located in the centre of the figure. Conservation of angular momentum implies that the images are always aligned with lens and source star.*

(Bottom panel reprinted with permission from Paczyński, B., Gravitational microlensing in the local group, *Annual Review of Astronomy & Astrophysics,* **34**, 419–459, 1996.)

With the angular Einstein radius (Einstein 1936)

$$\theta_E = \sqrt{2\,R_S\,\left(D_L^{-1} - D_S^{-1}\right)}\,,\tag{5}$$

one can define dimensionless angular coordinates on the sky

$$x = \frac{\xi}{D_L\,\theta_E}\,,\quad y = \frac{\eta}{D_S\,\theta_E}\,.\tag{6}$$

With these and the deflection law, Eq. (3), the lens equation simplifies to

$$y = x - \frac{1}{x}\,,\tag{7}$$

which means that the angular Einstein radius θ_E provides the unique scale of gravitational microlensing, whereas the mapping of an image position to the true source position is described by a simple equation involving dimensionless quantities.

Solving the lens equation for the images leads to a quadratic equation, which provides two solutions

$$x_\pm = \frac{1}{2}\left(y \pm \sqrt{y^2 + 4}\right).\tag{8}$$

2.2   Image distortion and magnification

From the observer's perspective, shown in the right panel of Figure 1, the conservation of angular momentum means that the images fall on a line through lens and source star.

The two images are on opposite sites of the lens star, and one of them is inside the Einstein circle, which is centered on the lens star and has the angular radius θ_E, while the other one is outside. As the source passes close to the lens star, its images move around, approaching the Einstein circle as the source approaches the lens. The separation between the images is at least $2\,\theta_E$, and close to that value whenever $y \le 1$.

As can be seen, the bending of light also causes a distortion of the images. While the lens equation, Eq. (7), provides a relation between the separations from the optical axis, i.e. the radial coordinate r in a polar coordinate system (r, φ) perpendicular to it, the polar angle φ is conserved. Therefore, the radial distortion is given by dx/dy, while the tangential distortion (related to $r\,d\varphi$) reduces to x/y.

For source stars in the Galactic bulge, $D_S \sim 8.5$ kpc and $D_L \sim 6.5$ kpc are characteristic values for source and lens distance, respectively. This implies an angular Einstein radius

$$\theta_E \sim 600\,(M/M_\odot)^{1/2}\,\mu\text{as}\,,\tag{9}$$

so that the images cannot be resolved with current optical telescopes.

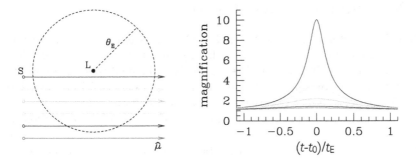

Figure 2. *Left: Trajectories (distinguished by colour) for a source star S, moving with a proper motion $\vec{\mu}$ with respect to the lens L, that pass at different impact parameter in relation to the angular Einstein radius θ_E, indicated by a dashed line. Right: The observed magnification as function of time t for the different trajectories, where the closest approach between lens and source star occurs at epoch t_0, and the event time-scale is given by $t_E = \theta_E/\mu$.*

However, the image distortion leads to an alteration of the receiving solid angle and therefore the observed flux. Since the magnification $A(y)$ is given by the absolute value of the Jacobian of the mapping, one obtains the simple analytic expression

$$A(y) = \sum_{\pm} \left| \frac{x_{\pm}}{y} \frac{dx_{\pm}}{dy} \right| = \frac{y^2 + 2}{y \sqrt{y^2 + 4}}, \qquad (10)$$

so that an observable brightening of the source star results.

2.3 Microlensing light curves

Since the magnification $A(y)$ is strictly decreasing with the lens-source separation y, microlensing light curves, as shown in Figure 2 along with the respective source trajectories, are symmetric around a peak, if one assumes a uniform proper motion $\vec{\mu}$ between lens and source star. For the same reason, the closer on the sky the source trajectory comes to the lens, the larger the peak magnification. The quotient of angular Einstein radius and proper motion defines an event time-scale $t_E = \theta_E/\mu$, where

$$t_E \sim 40 \, (M/M_\odot)^{1/2} \, \mathrm{d} \qquad (11)$$

for a typical proper motion $\mu \sim 15 \, \mu$as. With

$$y(t) = \sqrt{u_0^2 + \left(\frac{t - t_0}{t_E} \right)^2}, \qquad (12)$$

the magnification $A[y(t)]$ is determined by the three quantities u_0, t_0, and t_E, where $u_0 \, \theta_E$ is the minimal separation, realized at epoch t_0.

3 Microlensing event rate

As revealed by Renn, Sauer & Stachel (1979), Einstein's notes show that he discussed gravitational microlensing already in 1912, but it needed intense persuasion from a third party before he finally published these in 1936. However, he was still not convinced that his findings were useful and concluded 'There is no great chance of observing this phenomenon' (Einstein 1936).

Let us quantify that chance by considering an alignment between lens and source to be sufficient if a given source star is inside the Einstein circle of at least one of the lens stars. The probability for this to happen then defines the *optical depth* τ of microlensing. Since, with Eq. (10), such an alignment corresponds to a magnification of $A(u = 1) = 3/\sqrt{5} \approx 1.34$, the optical depth corresponds to the probability that a given star is magnified by more than 34 % at a given time.

With the optical depth τ being small, one can simply consider the sum of the solid angles of the sky covered by the Einstein circles of each of the N_L lens stars, given by $N_L \, \pi \, \theta_E^2$, and safely neglect any overlap. The number of lenses can be expressed by the volume mass density $\rho(D_L)$ and the mass spectrum $f(M)$, where $\int f(M) \, dM = 1$, integrated along the line-of-sight:

$$N_L \to \int f(M) \, D_L^2 \, \frac{\rho(D_L)}{M} \, dD_L \, d\Omega \, dM . \tag{13}$$

A rough estimate can be obtained for sources in the galactic bulge and lens stars in the galactic disk with the simplifying assumptions of both the distance to the source stars D_S and the mass density $\rho(D_L)$ of the lens stars being approximately constant for sources at a few degrees off the galactic plane. With $x = D_L/D_S$, inserting θ_E, as given by Eq. (5), and dividing by the total solid angle, yields the optical depth due to disk lenses

$$\tau_{\text{disk}} = \frac{N_L \, \pi \, \theta_E^2}{\int d\Omega} = \frac{4\pi G}{c^2} D_S \, \rho \int_0^1 x(1 - x) \, dx = \frac{2\pi G}{3c^2} D_S^2 \, \rho . \tag{14}$$

Inserting $D_S \sim 8$ kpc and a mass density $\rho \sim 0.1 \, M_\odot/(\text{pc})^3$ gives $\tau_{\text{disk}} \sim 6 \times 10^{-7}$. A more thorough analysis (Kiraga & Paczynski 1994) reveals that 65 % of the total optical depth towards the Galactic bulge arises from lensing by bulge stars rather than disk stars, so that $\tau_{\text{total}} \sim 2 \times 10^{-6}$. Very roughly, and easy to remember, of the order of one in a million stars in the Galactic bulge appears to be significantly brightened by microlensing at any given time.

This result translates to an event rate as follows. Mao & Paczyński (1991) distinguished 3 different characteristics of microlensing, where the set of source positions relative to the lens for which a feature lasts defines an area on the sky, the range of impact parameters of moving sources defines a width, and the parts of the source trajectories that are located inside the respective area define an average length. In particular, for a magnification exceeding 34 %, one finds an area of $\pi \, \theta_E^2$, a width of $2 \, \theta_E$, and an average length of $(\pi/2) \, \theta_E$.

The average duration of such an event is therefore

$$\langle t_e \rangle = \frac{\pi}{2} t_E \,, \tag{15}$$

where $t_E \sim 20$ d for $M \sim 0.3\, M_\odot$. The event rate Γ is then simply the quotient of optical depth τ and average event duration $\langle t_e \rangle$, multiplied by the number of observed source stars N_S, i.e.

$$\Gamma = N_S \frac{\tau}{\langle t_e \rangle} \sim 2 \times 10^{-5} N_S \,\mathrm{yr}^{-1} \,. \tag{16}$$

If one neglects the seasonal variations of target observability, 1000 observed events per year would therefore require (roughly daily) monitoring of 50 million stars, where about 85 events would be ongoing at any given time.

This means that microlensing surveys are a major venture in data processing. With no computers existing in 1936, Einstein could not imagine that one would eventually be able to carry out such a job successfully. Even around 1965, computers were still far too limited in power, and CCD cameras were not available either. It was in fact not until 1993 that the first microlensing event was reported (Alcock et al. 1993), resulting from observations by the MACHO team towards the Large Magellanic Cloud, rather than the Galactic Bulge. For these observations, the MACHO team used a telescope in Australia, and transferred all their data to the United States for analysis, which occupied about 30 % of the total available internet bandwidth between the two countries at that time. Nowadays, the OGLE and MOA surveys, using a dedicated 1.3m telescope at Las Campanas (Chile) and a 1.8m telescope at Mt John (New Zealand), respectively, monitor more than 100 million stars on a daily basis and announce close to 1000 microlensing events being in progress (Udalski et al. 1994, Bond et al. 2001).

4   Where to find planets

Another well-known technique for detecting planets exploits the fact that if a dark or dim object passes in front of a bright source object, the former eclipses part of the light of the latter and therefore creates a dimming. A recent spectacular occurrence of this phenomenon was the transit of Venus in front of the Sun in 2004.

But wait, we have also learnt that a foreground object bends the light of the background source object, making it to actually appear brighter. In fact, it depends on the geometry as well as the mass and size of the foreground object whether a brightening or a dimming results. One is tempted to think that for an eclipse to occur, the foreground object has to occult the background source object. However, this is not the exact condition. Instead, it has to occult the images that result from the bending of light by its gravitational field.

The dominance of one effect over the other depends on whether the angular radius of the foreground object θ or its angular Einstein radius θ_E is the larger.

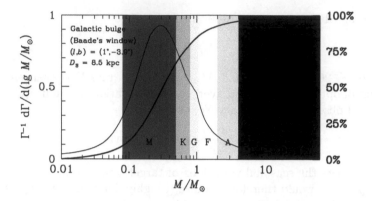

Figure 3. *Contribution of stellar and sub-stellar masses to the total event rate for a source located towards Baade's window with Galactic coordinates $(l, b) = (1°, -3.9°)$ at a distance of $D_S = 8.5$ kpc, where the adopted model of the Galaxy has been described by Dominik (2006). The thick line shows the cumulative distribution. Spectral types of stars are labelled O to M, where an indicative colour is shown for each.*

As the right panel of Figure 1 shows, for $\theta \ll \theta_E$, the foreground object will only manage to hit a very faint image for large lens-source separations, for which the lensing effect is close to negligible, whereas whenever lensing effects are substantial, the images are separated from the foreground lens star by about θ_E, and thereby escape occultation. Only if $\theta \ll \theta_E$, the foreground star occults images that result from gravitational bending of light.

In general, occultations occur if the foreground object is either close to the source object or close to the observer, whereas a brightening by microlensing is favoured for the foreground object being half-way between the observer and the source object. The relative width of the region for which microlensing occurs broadens for more distant sources. As a consequence, planets orbiting their observed star eclipse it, whereas microlensing only finds planets around the foreground lens stars. Moreover, stellar binaries lead to eclipses rather than lensing, and we see the Moon occulting the Sun during a solar eclipse.

Since the lens stars that host the planets detectable by microlensing are not observed themselves, there is no opportunity to select specific host stars, and very little opportunity to select the type of host star. Basically, the stellar and sub-stellar mass function determines the contribution of lens objects within any specific mass range to the ongoing microlensing events, as shown in Figure 3. Given that red dwarf stars are quite common, they make up about 50 % of the host stars of potential microlensing planets, whereas less than 20 % are expected to be of spectral type G or earlier.

5   Binary microlenses

In order to investigate the effect of planets around the lens star, we need to consider multiple lens objects. With coordinates $\vec{x}$ and $\vec{y}$ referring to the two-dimensional position angles of image and source star, scaled by the angular Einstein radius $\theta_{\rm E}$, the lens equation for a single point-mass lens, Eq. (7), takes the two-dimensional form

$$\vec{y} = \vec{x} - \frac{\vec{x}}{|\vec{x}|^2} \, . \tag{17}$$

For weak gravitational fields, the total deflection term results from superposition of the deflections independently caused by the individual lens objects; however, this does not mean that there is a superposition of light curves (the latter holds if the source is a multiple object). One can retain a form similar to that of Eq. (17) for the lens equation, if the mass M in the expression of the angular Einstein radius, Eq. (5), is chosen as the total mass. With m_i denoting the mass fraction of the i-th lens object, where $\sum m_i = 1$, and $\vec{x}^{(i)}$ being its coordinates, one finds

$$\vec{y} = \vec{x} - \sum m_i \, \frac{\vec{x} - \vec{x}^{(i)}}{\left|\vec{x} - \vec{x}^{(i)}\right|^2} \, . \tag{18}$$

For almost all cases, a planetary system can be approximated as just a binary lens composed of the host star and the dominant planet. A binary point-mass lens is completely characterized by two dimensionless parameters d and q, where $d\,\theta_E$ is the instantaneous angular separation, and $q = m_2/m_1$ is the planet-to-star mass ratio. Moreover, the time-scale of any effect caused by a planet is usually much smaller than the orbital period, so that the orbital motion can be neglected.

With $m_1 = 1/(1+q)$, $m_2 = q/(1+q)$, and, in a centre-of-mass system

$$\vec{x}^{(1)} = \left(\frac{q}{1+q}\, d, 0\right) \quad \vec{x}^{(2)} = -\left(\frac{1}{1+q}\, d, 0\right) , \tag{19}$$

the lens equation reads

$$y_1 \;=\; x_1 - \frac{1}{1+q}\, \frac{x_1 - \frac{q}{1+q}\, d}{\left(x_1 - \frac{q}{1+q}\, d\right)^2 + x_2^2} - \frac{q}{1+q}\, \frac{x_1 + \frac{1}{1+q}\, d}{\left(x_1 + \frac{1}{1+q}\, d\right)^2 + x_2^2} ,$$

$$y_2 \;=\; x_2 - \frac{1}{1+q}\, \frac{x_2}{\left(x_1 - \frac{q}{1+q}\, d\right)^2 + x_2^2} - \frac{q}{1+q}\, \frac{x_2}{\left(x_1 + \frac{1}{1+q}\, d\right)^2 + x_2^2} . \tag{20}$$

It leads to a 5th-order complex polynomial for the images (Witt & Mao 1995). The problem however can be reduced to a single master equation that is of 5th order in a real variable with real coefficients (Asada 2002). In any case,

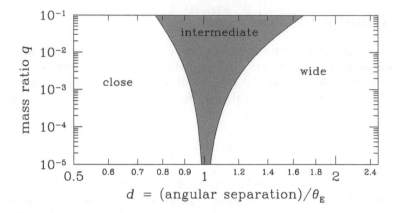

Figure 4. *The three different topologies of the caustics for a binary lens, depending on the mass ratio and the angular separation in units of the angular Einstein radius θ_E of the total mass. There is a 'resonance' around $d \sim 1$.*

numerical techniques are required in order to find solutions. Depending on the source position, there are either 3 or 5 images. As before, the magnification is given by the inverse absolute value of the Jacobian

$$A(\vec{y}) = \sum_j \left| \det \left(\frac{\partial \vec{y}}{\partial \vec{x}} \right) \left(\vec{x}^{(j)} \right) \right|^{-1} . \tag{21}$$

If the Jacobian vanishes for at least one image $\vec{x}^{(j)}$, the point-source magnification $A(\vec{y})$ tends to infinity. Such an image $\vec{x}^{(j)}$ is called *critical*, and the set of all source positions with $A(\vec{y}) \to \infty$, given by

$$\left\{ \vec{y}_c = \vec{y}(\vec{x}_c) \, \middle| \, \det \left(\frac{\partial \vec{y}}{\partial \vec{x}} \right) (\vec{x}_c) = 0 \right\} , \tag{22}$$

defines the *caustics* of the lens system.

For a point lens, critical points populate the Einstein circle $x = 1$, thereby forming a single closed curve, which is mapped onto the location of the lens $y = 0$, where a point-like caustic resides. For binary lenses, caustics form closed curves that adopt one of 3 possible topologies (Erdl & Schneider 1993), depending on the separation and mass ratio. As illustrated in Figure 4, there is a region with a single intermediate caustic around the 'resonant case', where the angular separation is about θ_E, i.e. $d = 1$, which narrows down towards smaller mass ratios. There are two separated caustics for the wide binary case, and three caustics for the close binary case.

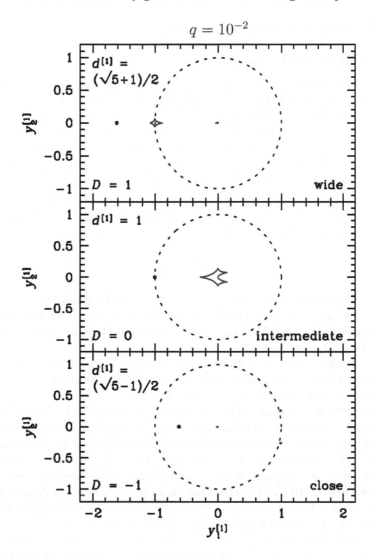

Figure 5. *Caustics for a binary lens with a mass ratio $q = 10^{-2}$, roughly corresponding to a planet of 3 Jupiter masses orbiting a star of $0.3\ M_\odot$, which illustrate the three different topologies. The star is located in the centre of the coordinate system and coordinates refer to angles in units of the angular Einstein radius of the stellar mass. While $d^{[1]}$ denotes the planet-star separation in these coordinates, D is the position for which bending of light by the star would produce an image at the angular location of the planet. For each of the panels shown, the planet is located along the horizontal axis, and its position is marked by a black filled circle. The chosen separations $d^{[1]}$ correspond to the two boundaries of the 'lensing zone' and the planet falling onto the Einstein circle of its host star (shown as a dashed line), respectively.*

6 Planetary microlensing

6.1 Images and caustics

For small mass ratios, we are dealing with a two-scale problem. Despite the fact that Chang & Refsdal (1979) had in mind the effect of individual stars on the images of quasars gravitationally lensed by galaxies, their discussion applies in complete analogy to planets orbiting lens stars. In the first instance, the lens star with its much bigger mass creates two images. A planet that happens to lie in the vicinity of either of these can then split it further. Let $d^{[1]} = d \sqrt{1+q}$ denote the angular planet-star separation in units of the stellar angular Einstein radius $\theta_E^{[1]} = \theta_E/\sqrt{1+q}$. According to the lens equation of a point-mass lens, Eq. (7), a source star at separation $D\,\theta_E$ is mapped to the angular position of the planet, if

$$D = d^{[1]} - \frac{1}{d^{[1]}} . \tag{23}$$

Consistently, extended *planetary caustics* are created around $D\,\theta_E$, whose size depends on the shear produced by the gradient of the gravitational field of the lens star and is therefore not only a function of the mass ratio q, but also of the planet-star distance $d^{[1]}$.

In contrast, a *central caustic* near the position of the lens star results from orbiting planets breaking the symmetry and causing the stellar critical curve to deviate slightly from the Einstein circle.

The topology and shape of the planetary caustics differ between the close-binary case, where two triangular-shaped caustics occur away from the planet-star axis, whereas there is a single diamond-shaped caustic on the planet-star axis for wide binaries. In the intermediate case, the planetary and central caustics merge into a single caustic. Figure 5 shows the three different caustic topologies for $q = 10^{-2}$, corresponding to a planet of about 3 Jupiter masses around a star with $M \sim 0.3\,M_\odot$. While $d^{[1]} = 1$, i.e. the planet falling onto the Einstein circle, corresponds to $D = 0$, the values $d^{[1]} = (\sqrt{5}+1)/2$ and $d^{[1]} = (\sqrt{5}-1)/2$ correspond to $D = 1$ or $D = -1$, respectively. Therefore, the planetary caustics are within the Einstein circle for a planet in the so-called *lensing zone* $(\sqrt{5}-1)/2 \le d^{[1]} \le (\sqrt{5}+1)/2$. Being related to the equation $y = x - 1/x$, these values represent the golden ratio; an example missing in the book by Mario Livio (2002).

6.2 Excess magnification

For the same planet-star separations as adopted for Figure 5, Figure 6 shows the fractional excess magnification caused by the presence of planets with mass ratios $q = 10^{-2}$, $q = 10^{-3}$ or $q = 10^{-4}$, where detectable effects arise from regions that are substantially larger than the caustics. For a wide binary, one has positive deviations next to the planetary caustic, surrounded by negative

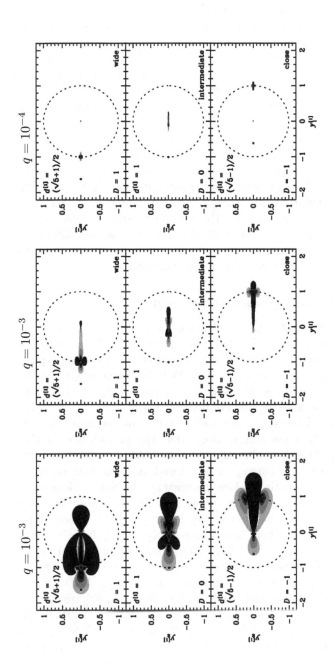

Figure 6. *Regions of the sky for which a planet orbiting the lens star creates a significant difference in magnification. While lighter shades correspond to a further brightening in presence of the planet, darker shades correspond to a dimming, where the shade levels refer to fractional excess magnifications of 1 %, 2 %, 5 % and 10 %. For comparison, the caustics are shown. From left to right, three different mass ratios have been chosen, namely $q = 10^{-2}$, $q = 10^{-3}$, and $q = 10^{-4}$, roughly corresponding to planets of 3 Jupiter masses, Saturn mass, and 10 Earth masses, respectively, for a stellar mass of 0.3 $M_\odot$. For each mass ratio, three different separations have been chosen, corresponding to the two boundaries of the 'lensing zone' and the resonant case where the planet happens to lie on the Einstein circle of the lens star, which illustrate the three different caustic topologies. The choice of coordinates is the same as in Fig. 5.*

deviations, whereas for the corresponding close binary, positive and negative deviations are roughly inverted, apart from the strong positive signals that surround the caustics. Intermediate separations $d^{[1]} \sim 1$ mark the transition.

If one compares the different mass ratios $q = 10^{-2}$, i.e. typically a planet 3 times more massive than Jupiter, $q = 10^{-3}$, roughly matching a Saturn-mass planet, and $q = 10^{-4}$, corresponding to a planet of about 10 Earth masses, one sees that the size of the deviation regions decreases, where the angular extent is roughly proportional to $\sqrt{q}$ near the planetary caustics, and proportional to q near the central caustic. For point-like source stars, both the signal duration and the probability scale with this factor, whereas the possible signal amplitude is only limited by the finite angular radius $\theta_\star$ of the observed source star. In fact, as the finite source sweeps over it, the deviation map is smoothed, and in particular small-angle structures containing positive and negative deviations are washed out. For sufficiently small mass ratios, such as $q = 10^{-4}$ in our examples, the deviation regions become clearly separated. The shear arising from the gradient of the gravitational field of the lens star increases the deviation regions and thereby increases the prospects for detecting planets, most substantially in the 'lensing zone'.

6.3   Light curves and planet detection

Light curves result from applying a one-dimensional cut through the two-dimensional magnification map. For the light curves with planetary signals shown in Figure 7, resulting from passages of the source trajectory near planetary caustics, an angular source radius $\theta_\star = 0.025\,\theta_E$ has been adopted, which roughly corresponds to a giant source star with a radius $R_\star \sim 15\,R_\odot$. For identical trajectories, less massive planets produce signals with both smaller amplitude and duration. However, it is possible to produce signals with large amplitudes even for small mass ratios, albeit that the probability for these to occur becomes tiny. Most prominently, the wide-binary cases show a rise, while a drop is observed for the corresponding close-binary configurations. As an exception to this rule, a caustic has been touched for the chosen close binary with $q = 10^{-4}$, resulting in a relatively strong positive deviation. Similarly, the source trajectory hit the wide-binary caustic for $q = 10^{-2}$, producing an extraordinarily strong signal. There is no comparably strong dimming in the corresponding close-binary case. For giant source stars, one finds durations of planetary signals that are not that short, namely ~ 2 days for 10 $M_\oplus$ and more than a week for a planet a few times more massive than Jupiter.

The finite-source effects are illustrated more explicitly in Figure 8, where the maximal signals for a main-sequence source star ($\theta_\star = 0.0017\,\theta_E$, i.e. $R_\star \sim 1\,R_\odot$) and a giant source star ($\theta_\star = 0.025\,\theta_E$, i.e. $R_\star \sim 15\,R_\odot$) passing directly over the planetary or the central caustic for a planet with mass ratio $q = 10^{-4}$ at the outer edge of the 'lensing zone' are compared. While a solar-type star passing over a planetary caustic of the 10 Earth-mass planet produces a huge signal (lasting about 5 h), it is smoothed out for a giant

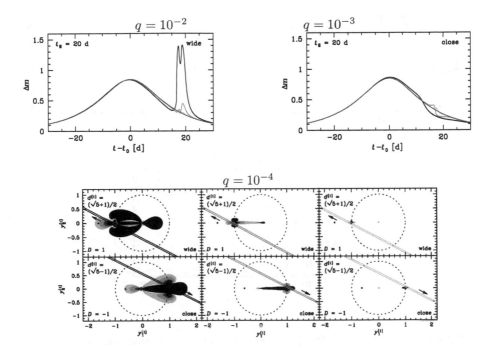

Figure 7. *Deviations resulting from a giant source star of radius $R_\star \sim 15 R_\odot$ passing in the vicinity of planetary caustics associated with a planet at the edge of the 'lensing zone'. While the left panel shows the excess brightening for a wide planet, the right panel shows the corresponding dips resulting from the presence of a close planet. Together with the shaded deviation regions shown already in Figure 6, the lower panel shows the chosen source trajectory. A 'typical' event time-scale of $t_E = 20$ d has been assumed.*

source star, thereby pushing the duration to more than a day, but remains at notable 20 %. The planetary deviations arising from the central-caustic passage suffer more substantially, where the signal for a star of solar radius does not exceed 7 %, while a planet would hardly be detectable for a giant source star.

While for a single point lens, the magnification does not depend on the orientation angle of the trajectory, only a fraction of the trajectory angles will lead to a detectable planetary signal. This implies that, for a given event with model parameters (u_0, t_0, t_E), there is a finite detection efficiency for planets at separation parameter d (or $d^{[1]}$, which is practically identical) and with mass ratio q, given by the probability

$$\varepsilon(d, q; u_0, t_0, t_E) = \qquad (24)$$
$$P\left[\text{detectable signal in event}(u_0, t_0, t_E)|\text{planet}(d, q)\right].$$

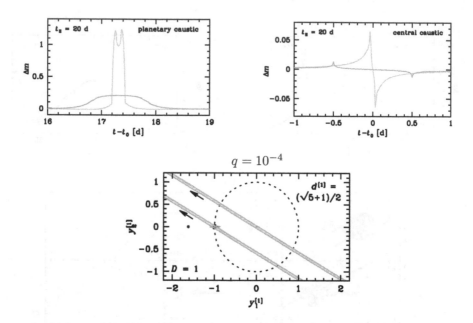

Figure 8. *Effects of the finite size of the source star on the planetary deviations, illustrated by the comparison of a main-sequence source star ($R_\star \sim 1\ R_\odot$) and a giant source star ($R_\star \sim 15\ R_\odot$) passing directly over the planetary or the central caustic for a planet with mass ratio $q = 10^{-4}$ at the outer edge of the 'lensing zone'. As before, an event time-scale of $t_E = 20$ d has been adopted.*

Since planets can be detected from the source star entering the vicinity of either the central or planetary caustics, two types of strategy can be distinguished. For sufficiently small u_0, central caustics are probed, so that potential deviations are found around a highly-magnified peak. Since a large range of angles produces a detectable signal, the detection efficiency is high; however, there are fewer events, the smaller the adopted upper limit on u_0. In particular towards less massive planets, the detection from approaches to planetary caustics harbours the larger potential, given the larger associated deviation regions that are less susceptible to a wash-out by finite-source effects. However, in order to make use of these theoretical prospects, a much larger effort is required, involving the monitoring of lots of events, each having rather modest detection efficiency, over a substantial fraction of their duration, since the planetary deviation could happen anytime.

References

Alcock, C. et al. (1993), Possible gravitational microlensing of a star in the large magellanic cloud, *Nature* **365**, 621.

Asada, H. (2002), Images for a binary gravitational lens from a single real algebraic equation, *Astron. Astrophys.* **390**, L11–L14.

Bond, I. A. et al. (2001), Real-time difference imaging analysis of MOA Galactic bulge observations during 2000, M*on. Not. R. Astron. Soc.* **327**, 868–880.

Chang, K. & Refsdal, S. (1979), Flux variations of QSO 0957+561 A, B and image splitting by stars near the light path, *Nature* **282**, 561– 564.

Dominik, M. (2006), Stochastic distributions of lens and source properties for observed galactic microlensing events, *Mon. Not. R. Astron. Soc.* **367**, 669–692.

Eddington, A. S. (1919), The total eclipse of 1919 May 29 and the influence of gravitation on light, *The Observatory* **42**, 119–122.

Eddington, A. S. et al. (1919), Discussion on the theory of relativity, *Mon. Not. R. Astron. Soc.* **80**, 96–118.

Einstein, A. (1911), Über den Einfluss der Schwerkraft auf die Ausbreitung des Lichtes, *Annalen der Physik* **340**, 898–908.

Einstein, A. (1915), Erklärung der Perihelbewegung des Merkur aus der allgemeinen Relativitätstheorie, *Sitzungsberichte der Königlich PreuSurface Scienceischen Akademie der Wissenschaften (Berlin)* **47**, 831–839.

Einstein, A. (1936), Lens-like action of a star by the deviation of light in the gravitational field, *Science* **84**, 506–507.

Erdl, H. & Schneider, P. (1993), Classification of the multiple deflection two point-mass gravitational lens models and application of catastrophe theory in lensing, *Astron. Astrophys.* **268**, 453–471.

Kiraga, M. & Paczynski, B. (1994), Gravitational microlensing of the galactic bulge stars, *Astrophys. J.* **430**, L101–L104.

Livio, M. (2002), *The Golden Ratio: The story of PHI, the world's most astonishing number*, Broadway Books, New York.

Mao, S. & Paczyński, B. (1991), Gravitational microlensing by double stars and planetary systems, *Astrophys. J.* **374**, L37–L40.

Renn, J., Sauer, T. & Stachel, J. (1979), 'The origin of gravitational lensing: A postscript to Einstein's 1936 Science paper', *Science* **275**, 184–186.

Udalski, A. et al. (1994), The optical gravitational lensing experiment. The early warning system: real time microlensing, *Acta Astronomica* **44**, 227–234.

Witt, H. J. & Mao, S. (1995), On the minimum magnification between caustic crossings for microlensing by binary and multiple stars, *Astrophys. J.* **447**, L105.

The practice of planet detection by gravitational microlensing: studying cool planets around low-mass stars

Martin Dominik

SUPA, University of St Andrews, School of Physics & Astronomy, North Haugh, St Andrews, KY16 9SS, United Kingdom

Royal Society University Research Fellow

Due to gravitational bending of light, a foreground 'lens' star passing close to the line-of-sight to an observed background star, causes an observable transient brightening, thereby creating a 'gravitational microlensing event'. The presence of a planet orbiting the lens star can lead to an additional short blip or dip lasting from hours to weeks. A lucky coincidence with respect to the typical values of the gravitational radius of stars and distances within the Milky Way gives us the opportunity to exploit this effect for detecting planets orbiting stars (preferentially the common K- and M-dwarfs) in both the Galactic disk and bulge at separations between 1 and 10 AU. With sensitivities reaching down to planets of Earth mass and even below, gravitational microlensing allows the probing of a region of planet parameter space that is hardly accessible by other techniques.

1 Gravitational microlensing events

Gravitational microlensing is understood as the bending of light received from an distant star due to the gravitational field of an intervening foreground star. This allows for two possible light rays from the source to the observer passing on either side of the so-called 'lens' star. These correspond to two distorted images, one inside and one outside the Einstein circle of radius

$$\theta_{\rm E} = \sqrt{2\,R_{\rm S}\,\left(D_{\rm L}^{-1} - D_{\rm S}^{-1}\right)}\,, \tag{1}$$

where $D_{\rm L}$ and $D_{\rm S}$ are the distances of the lens and source star, respectively, and $R_{\rm S} = (2GM)/c^2$ is the gravitational radius of the lens star, corresponding to its mass M. For typical distances within the Milky Way, the separation of less than a milli-arcsecond between the two images does not allow them to be resolved, whereas the alteration of the total received flux leads to a characteristic brightening on a time-scale of about a month, which increases with the proximity of alignment. Namely, for an angular separation $u\,\theta_{\rm E}$ between lens and source star, one finds a magnification (Einstein 1936)

$$A(u) = \frac{u^2 + 2}{u\,\sqrt{u^2 + 4}}\,. \tag{2}$$

A uniform proper motion μ between lens and source defines an event time-scale $t_{\rm E} = \theta_{\rm E}/\mu$, and since $A(u)$ is strictly decreasing with u, (ordinary) microlensing light curves are symmetric around a peak occuring at epoch t_0, where the lens-source angular separation is $u_0\,\theta_{\rm E}$, while the peak magnification $A_0 = A(u_0)$ increases with smaller u_0.

With only one in a million stars towards the Galactic bulge being magnified by more than 30 % at a given time, the OGLE and MOA surveys manage to monitor more than 100 million targets on a daily basis, and are therefore capable of providing the scientific community with about 1000 real-time alerts on ongoing microlensing events per year.

2   Microlensing planet searches

If a planet orbits the lens star, it can create an additional short blip or dip on the light curve (Mao & Paczyński 1991), lasting between hours and weeks, depending on the mass m of the planet, as well as the proper motion μ and the angular radius $\theta_\star$ of the source star. As long as the source stars can be approximated as point-like, planetary signals of arbitrary strength can occur, although both the probability for this to happen and the signal duration decrease with $\sqrt{q}$, where q denotes the planet-to-star mass ratio. Once the point-source signal duration drops below $(2\theta_\star)/\mu$, typically for $m \lesssim 10\ M_\oplus$, the finite angular size of the source star spreads the signal over a longer time, while its amplitude is reduced.

In order to detect and characterize these planetary signals, one needs to cater for a frequent round-the-clock coverage. This was first achieved by PLANET (Probing Lensing Anomalies NETwork), which started its operations with a one-month-long pilot campaign in 1995. Observing at 1-m class telescopes distributed in longitude around the southern hemisphere during the annual Galactic bulge observing season from May/June to August/September, the original goal of PLANET was to detect massive gas giant planets and to measure their abundance, whereas the detection of terrestrial planets was initially considered to be beyond the capabilities. PLANET aims at achieving a photometric precision of 1-2 % and a sampling between 1.5 and 2.5 h, so that deviations due to giant planets not only can be detected, but the planet can also be properly characterized from at least 10-15 data points during the course of such an anomaly. With given photometric precision and sampling interval, the required exposure time dictates the number of events that can be monitored. While for giant source stars, 20 events can be followed at a given time (or 75 per season), this reduces to 6 events at a given time and 20 per season for fainter, i.e. main-sequence stars (Dominik et al. 2002).

Gould & Loeb (1992) found that about 15 % of Jupiter-mass planets within the so-called lensing zone, comprising angular separations between 0.6 and 1.6 θ_E from their host stars, provide signals in excess of 5 % among all microlensing events where the source enters the Einstein circle of radius θ_E around the lens star, corresponding to a brightening by 34 %. However, Griest & Safizadeh (1998) stressed that the planet detection efficiency depends strongly on the event magnification, so that it reaches about 80 % if only events with $A_0 \gtrsim 10$ are considered. The detection of planets from deviations around the peaks of the most promising highly-magnified events (Rattenbury et al. 2002) is the goal of MicroFUN (Microlensing Follow-Up Network), who activate a telescope network on a target-of-opportunity basis. The bright targets involved allow the participation of amateur astronomers with access to 0.3-m telescopes.

In 1999, MACHO and OGLE-II provided about 100 alerts on ongoing microlensing events per year, among these 50 with $A_0 \geq 2$, 6 with $A_0 \geq 10$, and just 7 on giant source stars. As a consequence, PLANET was severely limited in its planet detection capabilities, so that rather than a maximum possible 75 events, only about 25 could have been monitored. This means that the number of expected annual detections of Jupiters roughly equalled 3 times their fractional abundance per host star. The OGLE-III upgrade in 2002 dramatically increased the number of available events (about 350 were issued in that year), and together with MOA, the mark of 1000 alerts per year is now being approached. This allows PLANET to approach its full theoretical capability of detecting 15-25 Jupiters times their fractional abundance per year (Dominik et al. 2002).

There are however roughly 4 times more ongoing events than PLANET can take data on with the desired precision and sampling rate. A larger number of events can be monitored with larger telescopes, such as the three UK-built robotic 2-m telescopes, the largest of their kind, that are currently exploited by RoboNet-1.0 (Burgdorf et al 2007). In 2005, RoboNet has joined forces with PLANET to form a common microlensing campaign. Robotic telescopes not only allow a fast response, but also a flexible scheduling. Moreover, it is not necessary to send observers to the telescopes, which currently binds a lot of financial and human resources of PLANET.

3   Detection efficiency and abundance limits

The planet affects the light curve only through two dimensionless parameters, which are the planet-to-star mass ratio q and the separation parameter d, where $d\,\theta_E$ is the instantaneous angular separation. For a given event, the planetary detection efficiency ε, i.e. the probability for a detectable signal to occur provided that a planet exists, is therefore a natural function of d and q. While the detection efficiency in general decreases towards smaller q, a resonance between the deflection angles due to the planet and its host star leads to separations around $d \sim 1$ being preferred, where the angular separation between planet and star equals the angular Einstein radius θ_E. It is a lucky coincidence that $R_S \sim$ few km, and the extent of the Milky Way with $D_S \sim 8.5$ kpc and $D_L \sim 6.5$ kpc combine to $D_L\,\theta_E \sim$ few AU, so that microlensing is sensitive to separations that correspond to typical planetary orbits, namely a range between 1 and 10 AU. Microlensing relies on chance alignments between source and lens stars, where the latter host the detectable planets. Given the stellar mass function, microlensing prefers low-mass stars.

In fact, the first abundance limit on planets orbiting M-dwarfs arose from PLANET observations on 42 well-covered events between 1995 and 1999 (Albrow et al. 2001), where less than a third (at 95 % confidence level) of the probed lens population, with typically $M \sim 0.3\ M_\odot$ were found to have Jupiter-mass companions in the range 1.5 AU $< a <$ 4 AU. The obtained limit corresponds to 9 expected planets (from essentially 3 seasons), whereas none had been observed. Snodgrass, Horne & Tsapras (2004) have also analyzed the planet detection capabilities of the first OGLE-III season, namely 2002. For the favourable separations, there is an average detection efficiency for Jupiter-mass planets of a few percent. However, given that the OGLE survey finds far more events, $\sim$600 per year now, than are monitored by PLANET, OGLE-III appears to be competitive as compared to PLANET for studying massive gas giants, whereas OGLE-II was not.

Over a wide range of planet masses, the detection efficiency for the PLANET/RoboNet campaign decreases with $\sqrt{q}$, whereas a sudden drop, closer to a proportionality with q, occurs for OGLE, due to the lack of sufficiently frequent coverage. Therefore, with Earth-mass planets only 20 times less likely to show up than their Jupiter-like counterparts, the large number of events provided by OGLE-III gave PLANET a chance to advance into the few-Earth-mass regime. The detection of such planets is also favoured by the fact that such planets have been predicted to be rather common, whereas gas-giant planets have meanwhile turned out to be rare around M-dwarfs.

4   The first planet detections

Observations by the OGLE and MOA surveys on microlensing event OGLE-2003-BLG-235/MOA-2003-BLG-53 led to the first undisputed claimed detec-

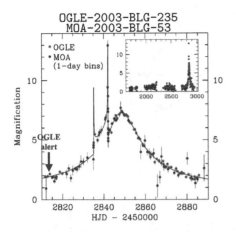
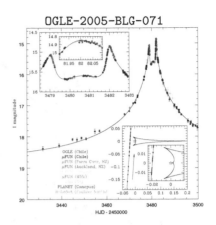

Figure 1. *(left) OGLE and MOA data along with a model light curve for event OGLE-2003-BLG-235/MOA-2003-BLG-53, revealing the first undisputed reported detection of a planet, with 1.5 Jupiter masses, by microlensing. An arrow indicates the time at which the OGLE collaboration first alerted the community about the ongoing event. (right) Data acquired by OGLE, Micro-FUN (μFUN), MOA, and PLANET/RoboNet at 9 different sites on OGLE-2005-BLG-071, as well as a model light curve, whose peak region shows the signal of a planet of about 3 Jupiter masses. The inset at the bottom right illustrates the approach of the source trajectory to the caustic located near the centre of the lens star, where the coordinates refer to angular positions in units of* θ_E *relative to the lens star.* (Left panel courtesy of David P. Bennett, right panel courtesy of Subo Dong & B. Scott Gaudi.)

tion of a planet, which turned out to be of about 1.5 Jupiter masses. In this case, the light curve, shown in the left panel of Figure 1, involves a quite strong signal with characteristic spikes, separated by about a week. These spikes result from the passage of the source star over a caustic, defined by the set of source positions for which the point-source magnification approaches infinity. Another planet with about 3 times the mass of Jupiter was found in event OGLE-2005-BLG-071 (Udalski et al. 2005), with a strong effect near the peak wit $A_0 \sim 45$. In this case, there is a model ambiguity between a wider and a closer planetary orbit. The corresponding light curve is shown in the right panel of Figure 1.

The most exciting discovery so far started its course on 11 July 2005, when the OGLE team alerted on their 390th event towards the Galactic bulge, on which a dense follow-up commenced 14 days later. The light curve of OGLE-2005-BLG-390 with data from 6 different sites is shown in the left panel of Figure 2. The event reached a peak at a magnification of 3 on 31 July, then faded, until a planetary deviation became first obvious with two points taken

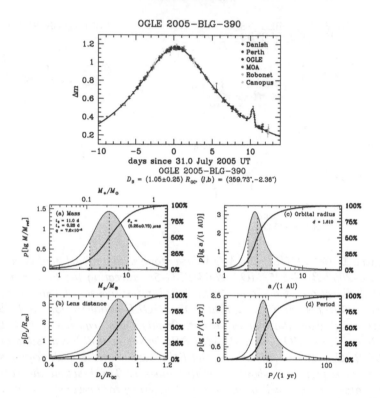

Figure 2. *(left) Data from 6 telescopes used by PLANET/RoboNet, OGLE, and MOA, along with a model light curve for event OGLE-2005-BLG-390. The observed deviation from an ordinary light curve over about a day revealed the presence of a planet of around 5 Earth masses. (right) Probability densities for the masses M_p and $M_\star$ of the planet and its host star, the orbital radius a and period P, assuming circular orbits, and their distance D_L from the observer in units of the distance to the Galactic centre R_GC. For the source distance, a prior distribution $D_\mathrm{S} = (1.05 \pm 0.25)\, R_\mathrm{GC}$ has been adopted. While the shaded regions include a probability of 68.3 %, half of it on each side of the median, the dot on the abscissa marks the expectation value, and the arrows show the standard deviation. The thick lines show the cumulated probability.*

by PLANET/RoboNet with the Danish 1.54m at ESO LaSilla. Subsequently, the falling part of this anomaly was followed by PLANET/RoboNet from Perth (Australia). The observed anomaly is manifested by an OGLE point from Chile and two points collected by MOA from New Zealand.

    Unfortunately, the light curve does not provide us with very much information. Besides the two parameters that characterize the planet (d and q), we can extract the event time-scale t_E, and the timespan $t_\star = \theta_\star/\mu$ during which the source moves by its own angular radius relative to the lens. With the angular radius θ_* of the source being determined from its luminosity and colour, pointing to a G4 III giant star with $R_* \sim 9.6\ R_\odot$ at $D_S = 8.5$ kpc, we find the lens-source proper motion $\mu = 7$ mas yr^{-1} and the angular Einstein radius $\theta_E = 210\ \mu$as. With these constraints, the probability densities shown in the right panel of Figure 2 resulted from the likelihood that given lens and source star properties gave rise to the observed event (Dominik 2006), where a mass function for the lens star, a distance for lens and source star, based on the spatial mass density of the Galaxy, and their velocity distribution have been assumed.

This gives us a 5 $M_\oplus$ planet, systematically named OGLE-2005-BLG-390Lb, around a 0.2 $M_\odot$ star at 3 AU orbiting within 10 years, everything within an uncertainty factor of two. The planet and its host star are about 15 % closer to us than the Galactic centre. Still within the uncertainty, the planet appears not to be massive enough to accrete a substantial amount of gas. With an estimated surface temperature of about 50 K, it should largely be made of ice, like a more massive version of Pluto, resembling the inner cores of Uranus or Neptune. Assuming a similar mass density, its radius would be 2.4 $R_\oplus$, and its surface gravity similar to that on Earth.

5 Probing planet parameter space

As shown in Figure 3, OGLE-2005-BLG-390Lb occupied a unique position amongst all exoplanets known at the time of its discovery, making it the closest analogue yet found to the rocky inner planets of the solar system. This reflects the unique capabilities of microlensing for detecting low-mass planets at several AU, and that it is still the only technique that can detect Earth-mass planets right now.

While the larger velocity and the larger range of inclinations for which an eclipse can occur favour closer orbits for the radial-velocity and the transit technique, respectively, the astrometric signal increases with orbit size. For all these techniques, a few orbits need to be observed, so that the duration of the campaign limits the orbit size for which planets can be detected. This restriction does not apply for microlensing, making it the only technique capable of detecting planets more distant than ~ 8 AU from their parent star within reasonable time, apart from a direct detection of emitted or reflected light from the planet. The microlensing limits are mostly of probabilistic nature, whereas the only principal limitation arises from the finite size of the source star washing out the signal for planets below a critical mass. Nevertheless, ground-based observations could reveal signals from planets even below Earth mass, and the higher photometric accuracy that can be achieved from space would make the detection of Martian planets likely.

The range of orbital radii for which conditions similar to Earth can exist that allow life forms to develop will be approached from the lower side by both the radial-velocity and transit searches. In contrast, microlensing probes the

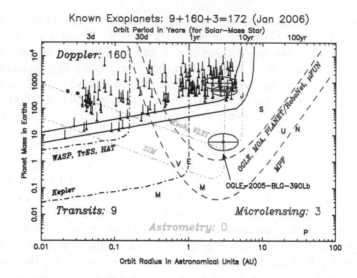

Figure 3. *Discovered exoplanets and detection limits from ground or space for various techniques as of January 2006 as function of planet mass and orbital radius or period (indicative): Radial velocity/Doppler wobble, astrometry, eclipsing transits, and microlensing. The planets of our Solar system are indicated by filled circles with a respective letter (M-V-E-M-J-S-U-N-P). The radial-velocity limits correspond to* 3 km s^{-1} *and* 1 km s^{-1}. *(Courtesy of Keith Horne.)*

abundance of planets on the other side of the habitable zone. The origin of habitable planets is however only understood once models of planet formation and orbital migration can be made to agree with observational data over a wide embracing region of parameter space that allows to distinguish between alternative explanations. With its complementarity to other techniques, microlensing observations will therefore be valuable for obtaining a complete picture.

6 Planetary census

In order to draw conclusions about the abundance of planets, one needs to compare the planet detections with the planet detection efficiency of the experiment, which gives the selection bias to the underlying population. The right panel of Figure 4 shows a preliminary plot of the average detection efficiency for the 14 most favourable events monitored by PLANET in 2004. While Jupiter-mass planets provide a detection with a probability up to 50 %, the detection efficiency roughly decreases with the square-root of the mass, until it drops off faster as the finite size of the source stars reduces it signif-

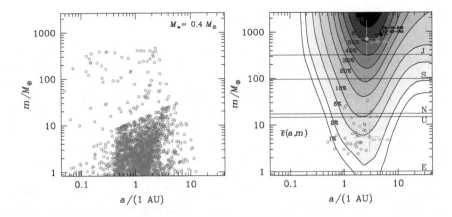

Figure 4. *(left) Distribution of planets with mass m and orbital radius a, resulting from a simulation based on a core-accretion model carried out by Ida & Lin (2005) for a stellar mass M = 0.4 $M_\odot$. (right) Average detection efficiency resulting from the 14 most favourable events monitored by PLANET in 2004, along with the microlensing planet detections and the distribution of detections expected from the planet distribution shown in the left panel. Horizontal lines mark the masses of Jupiter, Saturn, Neptune, Uranus, and Earth (J-S-N-U-E).*

icantly while a few Earth masses are being approached. However, this leaves detection efficiencies of a few percent for rocky and icy planets below 10 Earth masses.

In the PLANET sample, two planets showed up, OGLE-2005-BLG-071Lb with a mass a few times that of Jupiter and OGLE-2005-BLG-390Lb with a mass around 5 times that of Earth. The detection of the first microlensing gas giant planet by MOA and OGLE reflects their significant detection efficiency to such planets and an expected larger relative number of detections of heavier planets as compared to PLANET, but it cannot be properly compared with the detection efficiency of the PLANET campaign.

The left panel of Figure 4 shows the distribution of planets with mass and orbital radius for parent stars with M = 0.4 $M_\odot$, as resulting from a simulation of planet formation and orbital migration based on core-accretion models that have been carried out by Ida & Lin (2005). It predicts a quite small number of massive gas giants around M-dwarf stars, but a large number of rocky and icy planets between 0.4 and 3.5 AU. If this distribution is correct, the right panel of Figure 4 illustrates how the detections would be distributed. In order to reliably test models of planet formation and orbital migration, and to determine a first census of planets around K- and M-dwarfs, further detections are required.

7 Planets of Earth mass and below

While a planet of 5 $M_\oplus$ was easily detectable with a 15 % deviation over about a day in event OGLE-2005-BLG-390, an Earth-mass planet in the same spot would still have provided a 3 % signal, lasting approximately 12 h. If one assumes photometric uncertainties of $\sim$ 1 %, the discovery of such a planet would only have been possible if the standard follow-up sampling of 2 h had been replaced by high-cadence (10–15 min) anomaly monitoring triggered upon the first suspicion of a deviation. With real-time photometry and a prompt response from the telescopes, the SIGNALMEN anomaly detector (Dominik et al. 2007), which went into operation in 2007, is able to identify ongoing anomalies by successively requesting further observations until an anomaly can be confirmed or rejected with the required significance. Eliminating the effect of outliers by applying robust-fitting techniques, it provides initial triggers on data points whose absolute residual with respect to a current model light curve is within the largest 5 % of those of the full data set. The trade-off of this approach is in investing a fraction of the observing time into checking on potential anomalies for the return of the possibility to detect low-mass planets.

Figure 5 shows some worked examples for the RoboNet network, consisting of the Liverpool Telescope (LT), the Faulkes Telescope North (FTN), and the Faulkes Telescope South (FTS), augmented by two further identical telescopes in Chile and South Africa, for which the respective target observabilities were applied. For the respective simulation, an on-average $(2\,\text{h})/\sqrt{A}$ sampling, 1 % error at baseline and reduced according to photon statistics, as well as a further systematic error of 0.5 % have been assumed.

While several initial triggers would have led to the rejection of the anomaly hypothesis, all real anomalies would have been detected. The giant source star $(R_\star \sim 9.6\ R_\odot)$ that was observed in OGLE-2005-BLG-390 yielded a larger probability to detect a planetary signal and increased its duration, but reduced its amplitude as compared to a main-sequence star. Consistently, the signal duration for a main-sequence star $(R_\star \sim 1.2\ R_\odot)$ does not decrease proportionally to the source size. Provided that exposure times are chosen long enough for achieving a photometric accuracy of 1–2 %, the anomaly detector would also allow detection of an Earth-mass planet from a 5 % deviation in such a case. The detection of planets with masses as small as 0.1 $M_\oplus$ is challenging both by means of the short signal duration and the tiny probability for signals of appropriate amplitudes to occur, but possible in principle. In particular, the presence of telescopes in a location able to provide observations becomes crucial. In the right panel of Figure 5, a relevant part of the planetary anomaly happened to fall into a gap of just 40 minutes between the FTS (Australia) and South Africa. Among the chosen examples, this is the only one where the observations would not have been sufficient for properly characterizing the planet.

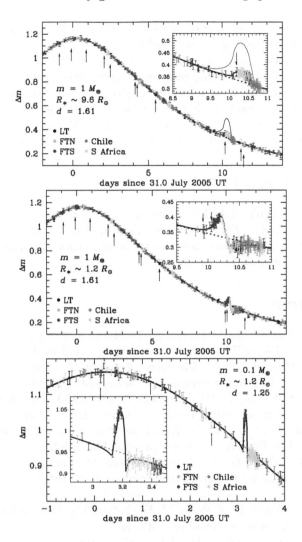

Figure 5. *Detection of planets of Earth mass and below, enabled by an anomaly detector with immediate feedback (simulation). Parameters not listed coincide with those of OGLE-2005-BLG-390 (see Figure 2). The arrows mark the selected initial trigger points, and the dashed line corresponds to the model light curve in absence of planets. (Top) The light curve shown as a thin line corresponds to OGLE-2005-BLG-390, involving a giant source star* ($R_\star \sim 9.6\ R_\odot$)*, whereas its planet has been replaced with an Earth-mass planet in the same spot for the main curve and the simulation. (Middle) As before, but with the source star now being 8 times smaller* ($R_\star \sim 1.2\ R_\odot$)*. Moreover, the trajectory angle has been adjusted in order to produce a 5% deviation. (Bottom) As before, but with a planet of* $m \sim 0.1\ M_\oplus$*, moved closer to its host star (*$d = 1.25$ *instead of the unfavourable* $d = 1.61$*). Again, the trajectory angle has been slightly adjusted in order to produce a significant deviation.*

A cooperative effort of current microlensing campaigns could provide detections of planets of Earth mass and below, using a coordinated three-step approach of survey, follow-up, and anomaly monitoring. Using real-time photometry from all microlensing campaigns using either robotic telescope networks or human observers, software carrying out real-time modelling, event prioritization, and anomaly detection could comprise an expert system that gives advice on optimal event monitoring; and with visualization and interpretation could even provide planet detection live to your home.

Acknowledgments

I would like to thank PLANET for providing preliminary results on detection efficiencies, as well as Shigeru Ida and Douglas Lin for providing the results of their core-accretion planet formation and migration simulations. Both sets of results were included in Figure 4.

References

Albrow, M. D. et al. (2001), Limits on the abundance of galactic planets from 5 years of PLANET observations, *Astrophys. J.* **556**, L113–L116.

Beaulieu, J.-P. et al. (2006), Discovery of a cool planet of 5.5 Earth masses through gravitational microlensing, *Nature* **439**, 437–440.

Bond, I. A. et al. (2004), OGLE 2003-BLG-235/MOA 2003-BLG-53: A planetary microlensing event, *Astrophys. J.* **606**, L155–L158.

Burgdorf, M. J. et al. (2007), Exoplanet detection via microlensing with RoboNet-1.0, *Planetary & Space Science* **55**, 582–588.

Dominik, M. (2006), Stochastic distributions of lens and source properties for observed galactic microlensing events, *Mon. Not. R. Astron. Soc.* **367**, 669–692.

Dominik, M., Horne, K. & Bode, M. F. (2006), The first cool rocky/icy exoplanet, *Astronomy & Geophysics* **47**, 25–3.

Dominik, M. et al. (2002), The PLANET microlensing follow-up network: results and prospects for the detection of extra-solar planets, Planetary & Space *Science* **50**, 299–307.

Dominik, M. et al. (2007), An anomaly detector with immediate feedback to hunt for planets of Earth mass and below by microlensing, *Mon. Not. R. Astron. Soc.* **380**, 792–804.

Einstein, A. (1936), Lens-like action of a star by the deviation of light in the gravitational field, *Science* **84**, 506–507.

Gould, A. & Loeb, A. (1992), Discovering planetary systems through gravitational microlenses, *Astrophys. J.* **396**, 104–114.

Griest, K. & Safizadeh, N. (1998), The use of high-magnification microlensing events in discovering extra-solar planets, *Astrophys. J.* **500**, 37–50.

Ida, S. & Lin, D. N. C. (2005), Toward a deterministic model of planetary formation. III. Mass distribution of short-period planets around stars of various masses, *Astrophys. J.* **626**, 1045–1060.

Mao, S. & Paczyński, B. (1991), Gravitational microlensing by double stars and planetary systems, *Astrophys. J.* **374**, L37–L40.

Rattenbury, N. J. et al. (2002), Planetary microlensing at high magnification, *Mon. Not. R. Astron. S.* **335**, 159–169.

Snodgrass, C., Horne, K. & Tsapras, Y. (2004), The abundance of Galactic planets from OGLE-III 2002 microlensing data, *Mon. Not. R. Astron. Soc.* **351**, 967–975.

Udalski, A. et al. (2005), A Jovian-mass planet in microlensing event OGLE-2005-BLG-071, *Astrophys. J.* **628**, L109–L112.

Modelling spectroscopic and polarimetric signatures of exoplanets

Daphne Stam

DEOS, Department of Aerospace Engineering, Technical University, Delft, and SRON Netherlands Institute for Space Research, Utrecht, the Netherlands

In order to find exoplanets that are not accessible with indirect detection methods and to characterize exoplanets, we have to detect and analyse the starlight that these planets reflect and/or the thermal radiation that they emit. In this chapter, we describe how planetary characteristics, such as atmospheric composition and structure, influence spectral features in the flux and degree of polarization of the planetary radiation, and what should be taken into account when modelling the planetary radiation.

1 Introduction

To this day, the majority of exoplanets is found using indirect detection methods, in which the planet itself is not detected, but rather its influence on its parent star. A famous method is the 'radial velocity method,' in which the velocity of the star due to the gravitational pull of an orbiting planet is measured by observing Doppler shifts in the starlight (Mayor & Queloz 1995). Another method is the 'transit method,' in which space telescopes like COROT (COnvection, ROtation & planetary Transits) stare at starfields in the sky to detect tiny drops in starlight that can be attributed to planets in orbits that lead them periodically across the stellar disk as seen from the telescope. Indirect detection methods are very strong tools for detecting exoplanets: at the time of writing, a decade after the detection of the first exoplanet around a solar-type star by Mayor & Queloz in 1995 (using the radial velocity method), the number of detected exoplanet candidates is more than 300.

Besides being useful for detections, the radial velocity method yields some physical parameters of the planetary system, such as the planet's orbital period (and, using an estimate of the star's mass, also the radius of the orbit), the ellipticity of the orbit, and a minimum limit on the planet's mass. The actual mass of the planet depends on the orientation of the planetary orbit, which cannot be derived from these observations. The transit method yields the planet's orbital period (and orbital radius), too, and from the drop in stellar flux during transit, the size of the planet can be estimated. In the rare cases in which a planet can be observed both using the radial velocity and the transit method, the actual mass of the planet and its density can be derived (for the first example of this, see Charbonneau et al. 2000). Careful measurements of the flux of transiting planetary systems during a transit can also give information about the composition of the upper atmospheric layers of the planet, because the observed starlight includes light that has traveled through these upper layers and that carries the absorption signatures of atmospheric constituents in these layers (for the first and a recent detection of an exoplanetary atmosphere, see Charbonneau et al. 2002; Swain et al. 2008).

Although indirect detection methods have provided and will provide a wealth of exoplanets and information about physical characteristics of a number of these planets, a thorough characterization of exoplanets requires the direct detection of planetary radiation. Planetary radiation consists of starlight that is reflected by the planet and of thermal radiation that is emitted by the planet, and carries information about the composition and structure of the planetary atmosphere and of the surface below the atmosphere, if there is any. This information will allow us to compare planets around other stars with the solar system planets, to better understand the origins and future of our planetary system, and to learn more about whether our system, with us in it, is special or very ordinary. Another compelling reason for direct detections is that the currently used indirect methods are most sensitive to large planets in tight orbits around their stars. Small planets and planets in wide orbits are difficult or simply impossible to find, unless one searches for the planetary radiation.

Direct observations of exoplanets are extremely difficult because planets are very dim compared to their parent stars (see Figure 1), and because the angular distance between an exoplanet and its star is very small. The typical angle $\delta = D/d$ arcsec, with the orbital radius D in Astronomical Units (AU) and the distance between the planet and the observer d in parsec (pc). For an Earth ($D = 1$ AU) at 10 pc, $\delta = 0.1''$, and for a Jupiter ($D = 5$ AU) at 10 pc, $\delta = 0.5''$. There have been some very interesting observations of planetary radiation derived from carefully observing a transiting planetary system just before and after the planet is in front of or behind its star, thus without spatially resolving the planet. A result of such observations was the first temperature map of an exoplanet, gas giant HD189733b (Knutson et al. 2007).

Spatially resolving a planet from its star requires special techniques to get rid of (most of) the starlight, and to enhance the contrast between a star and its planet, such as adaptive optics and coronographs. Using such techniques, some observations of spatially resolved planets around Brown Dwarfs, which are much dimmer than solar type stars, have been reported (see e.g. Chauvin et al. 2004). And, a first observation of a relatively young gaseous planet around a solar type star was announced in autumn 2008 (Lafreniére, Jayawardhana & van Kerkwijk 2008).

Another technique for enhancing the contrast between a star and a planet, is polarimetry. The reason for this is that, when integrated over the stellar disk, the direct light of a solar type star can be considered to be unpolarized (Kemp et al. 1987), while the starlight that has been reflected by a planet will usually be polarized, because it has been scattered within the planetary atmosphere and/or because it has been reflected by the surface (if there is any). The polarization signature will also help to determine whether or not the observed object is a planet or a background object. Besides detecting exoplanets, polarimetry can also be used for characterizing exoplanets, because the planet's degree of polarization as a function of wavelength and/or planetary phase angle is sensitive to the structure and composition of the planetary atmosphere and surface. This application of polarimetry is well-known from remote sensing of solar-system planets, in particular Venus (for a classic example, see Hansen & Hovenier 1974).

Several instruments are being designed and built for the direct detection of exoplanets. An example is the Spectro-Polarimetric High-contrast Exoplanet REsearch (SPHERE) instrument (Beuzit et al. 2006) for the Very Large Telescope (VLT) of the European Southern Observatory (ESO), which uses a coronograph, adaptive optics, differential imaging and polarimetry to detect and characterize gaseous exoplanets.

In this chapter we will concentrate on the numerical simulation of flux and polarization signals of terrestrial and gaseous exoplanets. Such simulations are essential tools for the development of instruments for direct detection for both ground-based and space-based telescopes (such as the filter selection and required accuracies), for the development of observing strategies (such as integration times), and of course for the interpretation of observed signals. In Section 2, we will describe properties of planetary radiation, optical properties of planets, and will explain how to calculate planetary signals. In Section 3, we will show flux and polarization spectra of the Earth, and discuss the origin and diagnostics of the spectral features, and we will show numerically simulated spectra for Earth-like and gaseous exoplanets. Section 4, finally, contains a short summary.

2   Describing and calculating planetary radiation

2.1   Flux and polarization

The flux and state of polarization of planetary radiation with wavelength λ can fully be described by a flux (column) vector $\mathbf{F}$ as follows (Hovenier et al. 2004; Hovenier & van der Mee 1983)

$$\mathbf{F}(\lambda) = [F(\lambda), Q(\lambda), U(\lambda), V(\lambda)]. \tag{1}$$

Here, F is the total flux of the radiation, Q and U describe the linearly polarized flux, and V the circularly polarized flux. Units of F, Q, U, and V are W m^{-2}. Following Eq. (1), unpolarized light is described by the (column) vector $\mathbf{F} = [F, 0, 0, 0] = F[1, 0, 0, 0] = F\mathbf{1}$, with $\mathbf{1}$ the unit (column) vector. A typical example of unpolarized light that will be used in this chapter, is light of a solar-type star that is integrated over the stellar disk (Kemp et al. 1987).

The polarized fluxes Q and U are defined with respect to a reference plane. We use the planetary scattering plane as the reference plane. This is the plane through the centers of the planet and the star that also contains the observer. Flux Q is the flux polarized parallel to the reference plane minus the flux polarized perpendicular to the reference plane, while U is the difference between the fluxes polarized under angles of 45° and 135° with the reference plane (for details, see e.g. Hansen & Travis 1974). The degree of polarization P of the radiation described by Eq. (1) is defined as

$$P(\lambda) = \frac{\sqrt{Q^2(\lambda) + U^2(\lambda) + V^2(\lambda)}}{F(\lambda)}. \tag{2}$$

P is independent of the choice of reference plane. Assuming that the planet is mirror-symmetric with respect to the reference plane, and that the incoming stellar light is unpolarized, polarized fluxes U and V of the disk-integrated planetary radiation will equal zero because of symmetry (remember that for years to come, direct detections of exoplanets will yield little more than disk-integrated signals). In that case, we can use an alternative definition of the degree of polarization, namely

$$P(\lambda) = -\frac{Q(\lambda)}{F(\lambda)}. \tag{3}$$

For $P > 0$ (i.e. $Q < 0$), the radiation is polarized perpendicular to the reference plane, thus perpendicular to the line the observer can draw between the centers of the planet and the star, while for $P < 0$ (i.e. $Q > 0$), the radiation is polarized parallel to the reference plane.

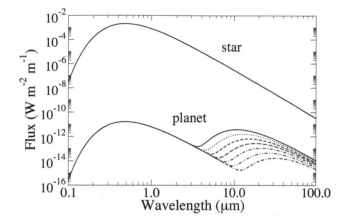

Figure 1. *The flux of a solar-type star, with an effective temperature of 6000 K and a radius of 700.000 km, and the planetary flux of a Jupiter-like planet, with a 70.000 km radius, a geometric albedo of 1.0, orbiting the star at 5.2 AU. Different lines represent different effective temperatures of the planet, in order: 268.8 K (solid line), 225.6 K, 189.3 K, 159.4 K, 133.2 K, 110.4 K, and 91.2 K (dot-dash-dash line).*

2.2   Planetary radiation

Given a spatially unresolved planet, the planetary flux vector, **F**, that arrives at the observer is simply the sum of the thermal radiation that is emitted by the planet, **F**$_t$, and the flux vector of the starlight that is reflected by the planet, **F**$_r$. Figure 1 shows, schematically, the flux of a solar-type star and the planetary flux of Jupiter-like, gaseous planets at various effective temperatures (Stam, Hovenier & Waters 2004). In each of the planet curves, the thermal flux is the bump at the longest wavelengths (above about 3 μm), and the reflected flux is the bump at the short wavelengths, with the same shape as the stellar flux (because it is reflected starlight). When calculating the curves in this figure, it was assumed that the star and the planet both radiate like black bodies, and spectral features, e.g. due to absorbing gases and/or absorbing and scattering atmospheric particles (see Section 2.1) are ignored. It is assumed that the disk of the planet is seen fully illuminated, i.e. the planetary phase angle α, which is defined as the angle between the star and the observer as seen from the center of the planet, is 0° (see Figure 2). In practice, a planet at this phase angle cannot be observed, because it will be located behind the star as seen from the observer. In Figure 1 and in the rest of this chapter, we ignore any influence of the Earth's atmosphere on the observed planetary radiation, such as absorption by terrestrial gases. Our observer is thus, for example, a space telescope.

The curves in Figure 1 clearly illustrate the huge contrast ratio between the fluxes of a star and its planet: in the visible, where the planetary flux consists of reflected starlight, the ratio is on the order of 10^8, while in the infrared, where the planetary flux is thermal radiation, it is about 10^4 for a hot planet (near $\lambda \approx 10 \ \mu$m), and about 10^2 for a cold planet such as Jupiter (near $\lambda \approx 40 \ \mu$m). Obviously, a planet's thermal flux depends strongly on the planet's temperature (see the curves in Figure 1). This temperature is strongly related to the age of the planet: with increasing age, the temperature usually decreases. The temperatures in Figure 1 have been taken from and are representative for planets with ages of, respectively, 0.125, 0.25, 0.5, 1.0, 2.0, 4.0, and 8.0 Gyrs (see Burrows, Sudarsky & Lunine 2003). Note that the planets in the solar system are about 4.5 Gyrs ($4.5 \cdot 10^9$ years) old.

The thermal flux arising from within the planet can be considered to be phase angle independent. Some planets will also emit a phase angle dependent thermal flux. Indeed, when a planet orbits its star in a tight orbit, the stellar flux will heat up the planet on the side that is facing the star (that's why close-in gaseous exoplanets are sometimes called hot Jupiters). This thermal flux and its phase angle dependence will depend on the planet's rotation rate (a fast rotating planet will heat up on all sides), the presence of an atmosphere (the surface of a rocky planet with a thin atmosphere will cool down rapidly when its star 'sets'), the thickness of the atmosphere (the thicker the atmosphere, the better the heat will be retained, and the smaller the phase angle dependence will be, even for a slowly-rotating planet), the dynamical processes within the atmosphere (winds can redistribute thermal energy across the planet), and of course on the stellar flux that is incident on the planet (which depends on the flux pouring out of the star, and on the distance between the planet and the star).

Given a planet's temperature, the thermal flux that can be observed from Earth is proportional to r^2/d^2, with r the radius of the (spherical) planet, and d its distance from Earth (see Figure 2). In Figure 1, we didn't include any spectral features in the thermal flux. In reality, the flux will show spectral features that are due to absorption and subsequent reemission of radiation by atmospheric constituents, such as gases and cloud particles. When the temperature decreases with altitude in the atmosphere, the features will appear as absorption bands (because in the deepest parts of the bands, the thermal radiation is emitted in the highest, and coldest, atmospheric layers). When the temperature increases with altitude, the features appear as emission bands.

Figure 1 only shows the total flux of the radiation, not its state of polarization. Like the stellar flux integrated over the stellar disk (Kemp et al. 1987), the planet's thermal flux integrated over the planetary disk will usually be unpolarized. In theory, it could, however, get polarized when it is scattered, for example by cloud particles. To gain insight into the degrees of polarization that could arise from such scatterings further numerical simulations and possibly observations of e.g. solar system planets, are required.

A planet's reflected flux that can be observed from Earth, is proportional to the stellar flux that is incident on the planet, which depends on the flux pouring out of the star, and on the distance between the planet and the star.

It is proportional to r^2/d^2, with r the radius of the (spherical) planet, and d its distance from Earth, and on the reflective properties of the planet (more details are given in the next section). To calculate the reflected flux curve in Figure 1, we assumed the planet reflects all the stellar flux that is incident on it isotropically in all directions (it has a geometric albedo of 1.0). In reality, this albedo will vary with wavelength and the reflection of the light will usually not be isotropic, depending on the composition and structure of the planetary atmosphere (see Section 3.2).

Finally, the flux of the reflected starlight will vary strongly with the fraction of the planetary disk that is illuminated and in view, and hence with the planetary phase angle α. The range of phase angles an exoplanet can in principle be observed at depends on the orientation of the orbit with respect to the observer: $90° - i \leq \alpha \leq 90° + i$, with i the orbital inclination angle. When $i = 0°$, the orbit is viewed face-on, and the planet can only be observed at $\alpha = 90°$. This phase, where half of the planetary disk is illuminated, is usually referred to as quadrature. When $i = 90°$, the orbit is viewed edge-on, and the planet can be observed at phase angles ranging from almost $0°$ to almost $180°$ (depending on the planet's orbital period; you might have to wait for decades to actually observe the planet at a range of phase angles). When $\alpha \approx 0°$ or $\alpha \approx 180°$, it will be impossible to spatially separate the extra-solar planet from its star and thus to measure F and/or P of the reflected starlight without including the direct stellar light. In particular, when $\alpha = 0°$, the planet is precisely behind its star, and when $\alpha \approx 180°$, the planet is transiting the stellar disk, and one might observe starlight that has been transmitted through the outer layers of the planetary atmosphere (Charbonneau et al. 2002, Swain, Vasisht & Tinetti 2008)

As seen from the Earth, outer solar system planets like Mars and the gaseous planets can only be observed at a limited range of phase angles (the maximum phase angle that Mars can be observed at is about $45°$, for Jupiter it is about $11°$, and for Saturn, $6°$). An inner planet like Venus, however, can be observed at almost all phase angles, except when α is around $0°$ (when we would see a fully illuminated Venus disk), because then Venus is located precisely behind the Sun as seen from Earth. When α is around $180°$, Venus can be observed as a black dot transiting the solar disk. For us, on Earth, Venus is thus observable as if it were an exoplanet.

Unlike the stellar light and the planet's thermal radiation, the starlight that is reflected by a planet will usually be polarized, because the light has been scattered by atmospheric particles, and because it has been reflected by the surface beneath the atmosphere, if there is any. The degree of polarization that can be measured when observing an exoplanet can thus be written as

$$P(\lambda) = -\frac{Q_{\mathrm{r}}(\lambda)}{F_{\mathrm{r}}(\lambda) + F_{\mathrm{t}}(\lambda) + F_0(\lambda)}. \tag{4}$$

Here, F_0 is the (unpolarized) stellar flux. Looking at the curves in Figure 1, it is clear that at infrared wavelengths up from about 10 μm, Q_{r} and F_{r} can be ignored, hence P will be zero. At visible and near infrared wavelengths up to about 2 μm, F_t can be ignored. In case the exoplanet is spatially unresolved from its star, for example, because it is orbiting its star in a tight orbit, the observable degree of polarization will be extremely small (on the order of 10^{-8}, or 0.000001 %), because of the huge contribution of the unpolarized stellar flux, F_0, to the total signal (for sample calculations, see Seager, Whitney & Sasselov 2000).

In the rest of this chapter, we assume that the radiation of the exoplanet can be observed without any interference of the stellar flux, in which case, $P = -Q_r/F_r$ (cf. Eq. (2)). In reality, this will be difficult, even for an exoplanet in a wide orbit (several AU) around its star and at intermediate phase angles, because the telescope optics will smear out at least some of the huge stellar flux across the image. Special observing tools, such as a coronograph (basically, a shade that is placed in or in front of the telescope to intercept as much as possible of the direct stellar flux) and, for ground-based telescopes, adaptive optics (flexible telescope mirrors with shapes that are constantly and automatically adapted to counteract the blurring effects of the atmosphere on the image), help to minimize F_0 and to bring out F_r and Q_r.

2.3   Reflected starlight

In this section, we discuss in more detail the influences on flux vector $\mathbf{F}_r$, the starlight that is reflected by a planet. Assuming a planet with radius r at a distance D from its star with radius R, and at a distance d from the observer (see Figure 2), this vector can be described by (Stam et al. 2006)

$$\mathbf{F}_r(\lambda, \alpha) = \frac{r^2}{d^2} \frac{R^2}{D^2} \frac{1}{4} \mathbf{S}(\lambda, \alpha) \, \pi \mathbf{B}_0(\lambda). \tag{5}$$

Here, we have introduced the dependence on phase angle α. Furthermore, $\mathbf{S}$ is the so-called planetary scattering matrix (see below), and $\mathbf{B}_0 = [B_0, 0, 0, 0] = B_0 [1, 0, 0, 0]$ is the flux (column) vector of the (unpolarized) stellar light, with πB_0 the stellar surface flux (in W m^{-2} m^{-1}). The stellar light is assumed to be unidirectional when it arrives at the planet (thus, $D \gg R$).

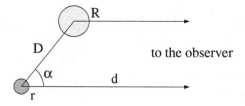

Figure 2. *Sketch of the geometries, with α the planetary phase angle, D the distance between star and planet, d the distance between planet and observer, R the stellar radius, and r the planetary radius.*

The planetary scattering matrix **S** describes the starlight that has been scattered within the planetary atmosphere and that has been reflected by the surface, if there is any, towards the observer (it does not contain diffracted starlight, such as that one might observe when $\alpha \approx 180°$). Matrix **S** refers to the reflected starlight integrated over the part of the planetary disk that is illuminated by the star and visible for the observer, because for years to come, exoplanets will be observable as point sources only. Assuming the planet is mirror-symmetric with respect to the planetary scattering plane (note that the long integration times that are required for observing an exoplanet will smear out many spatial features and yield a symmetric planet), matrix **S** has the following shape

$$
\mathbf{S}(\lambda, \alpha) = \begin{bmatrix} a_1(\lambda, \alpha) & b_1(\lambda, \alpha) & 0 & 0 \\ b_1(\lambda, \alpha) & a_2(\lambda, \alpha) & 0 & 0 \\ 0 & 0 & a_3(\lambda, \alpha) & b_2(\lambda, \alpha) \\ 0 & 0 & -b_2(\lambda, \alpha) & a_4(\lambda, \alpha) \end{bmatrix}. \tag{6}
$$

The matrix elements of **S** depend on the composition and structure of the planetary atmosphere and on the optical properties of the underlying surface. These elements hold the keys to the physical properties of a planet; once (some of) these elements have been measured, we can start characterizing a planet.

In particular, matrix element a_1 equals $4A_G$, with A_G the geometric albedo of the planet, which is defined as the ratio of the flux that is reflected by the planet at opposition (when $\alpha = 0°$) to the flux that is reflected by a Lambertian surface (i.e. a surface that reflects all incoming radiation isotropically and completely depolarized) that receives the same incoming flux and that subtends the same solid angle (i.e. $\pi r^2 / d^2$) on the sky. Integrated over the spectrum, the geometric albedo of Jupiter is about 0.5, that of Venus about 0.7 (high, because of Venus' highly reflecting cloud layer) and that of Earth about 0.4 (relatively low, because of the dark oceans).

In practice, with unpolarized incoming starlight, we can only measure two elements of matrix **S**, a_1 and b_1: the flux of the reflected starlight is given by

$$
F_r(\lambda, \alpha) = \frac{r^2}{d^2} \frac{R^2}{D^2} \frac{1}{4} a_1(\lambda, \alpha) \, \pi B_0(\lambda), \tag{7}
$$

and the degree of polarization by (see also Eq. (3))

$$
P(\lambda, \alpha) = -\frac{b_1(\lambda, \alpha)}{a_1(\lambda, \alpha)} \tag{8}
$$

The degree of polarization is thus independent of the sizes of the star and the planet, and of the distances between the star and the planet and between the planet and the observer. This is convenient, because accurate values of these parameters are not always available. Polarization P is also independent of the incoming stellar flux (although of course the number of photons that can be observed does determine the accuracy with which P can be measured) and its wavelength dependence (apart from Fraunhofer lines in the stellar spectrum; the depth of such high-spectral resolution features can change upon inelastic scattering by gaseous molecules, see Section 3.1).

2.4   Radiative transfer parameters

In this section, we'll describe which parameters influence the transfer of star-light within a planetary atmosphere, and with that the scattering matrix elements a_1 and b_1. Information about these parameters is embedded in the flux and state of polarization of the starlight that is reflected by an exoplanet.

We model the planetary atmosphere as a stack of homogeneous layers. Vertical inhomogeneities can be taken into account by choosing an appropriate number of different layers. When the planet is of the terrestrial, i.e. rocky, type, the atmosphere is bounded below by a surface that absorbs and/or reflects radiation. The atmospheric layers contain gaseous molecules, and, optionally, aerosol particles. Aerosol is the generic name for small airborne particles, such as mineral particles and photochemical haze particles. Aerosol sizes range from 0.01 μm to several tens of microns in diameter. In the following, we will also regard cloud particles, which have typical diameters of a few tens of microns, as aerosol particles. Solid particles, such as dust and ice cloud particles, will usually have irregular instead of spherical shapes (in that case, the diameters mentioned are those of spheres that have the same volumes as the irregular particles).

To understand the transfer of radiation through the atmosphere, you have to know what fraction of the radiation that is incident on the top and the bottom of a layer is intercepted within the layer, what fraction of the intercepted radiation is scattered (the rest is absorbed), what the angular distribution of the scattered radiation looks like, and what the reflection properties of the surface are like. This is described by, respectively, the layer's extinction optical thickness, the albedo of the particles in the layer, the single scattering matrix of these particles, and the albedo and reflection matrix of the surface details and more references can be found in Stam (2008).

2.4.1   Optical thickness

The extinction optical thickness b^{ext} of an atmospheric layer (which is usually referred to as 'the optical thickness') describes the extinction of the flux F_0 of an incident monodirectional beam of light through the layer

$$F_0'(\lambda) = F_0(\lambda)\mathrm{e}^{-b^{\text{ext}}(\lambda)/\cos\theta_0}. \tag{9}$$

Here, θ_0 is the angle of incidence of the beam of light, measured from the upward vertical ($0° \leq \theta_0 < 90°$). The flux $F_0 - F_0'$ has disappeared from the direct beam of light because of absorption and/or scattering by the particles in the atmospheric layer. The extinction optical thickness of the atmospheric layer is the sum of the absorption and the scattering optical thicknesses, as follows

$$b^{\text{ext}}(\lambda) = b^{\text{abs}}(\lambda) + b^{\text{sca}}(\lambda) = \sigma^{\text{abs}}(\lambda)n + \sigma^{\text{sca}}(\lambda)n = \sigma^{\text{ext}}(\lambda)n, \qquad (10)$$

with σ^{ext}, σ^{abs}, and σ^{sca}, the extinction, absorption, and scattering cross-sections (in m^2) of the particles, and n their column number density (in m^{-2}, i.e. this is the number of particles in the layer above each m^2). In case there are different types of particles in an atmospheric layer, such as both gaseous molecules and aerosol particles, their optical thicknesses can simply be added to get the total optical thicknesses.

The molecular scattering optical thickness of an atmospheric layer is given by

$$\begin{aligned}
b^{\text{sca}}_{\text{m}}(\lambda) &= \sigma^{\text{sca}}_{\text{m}}(\lambda)n_{\text{m}} \\
&= \frac{24\pi^3}{\lambda^4 N_{\text{L}}^2} \frac{(n_{\text{i}}^2(\lambda) - 1)^2}{(n_{\text{i}}^2(\lambda) + 2)^2} \frac{(6 + 3\rho(\lambda))}{(6 - 7\rho(\lambda))} \frac{N_{\text{av}}}{R} \int_{\Delta z} \frac{p(z)}{T(z)} dz. \qquad (11)
\end{aligned}$$

Here, the molecular column number density, n_{m}, is given by the last two terms, with z the altitude, Δz the geometrical thickness of the layer, p the pressure, T the temperature, N_{av} Avogadro's number, and R the gas constant per mole. Furthermore, N_{L} is Loschmidt's number, n_{i} is the refractive index of the planet's air under standard conditions, and ρ is the depolarization factor of the gaseous molecules. For molecules that are isotropic (in 'shape'), ρ equals zero. For non-isotropic gases, typical values for ρ are 0.02 for H_2, 0.09 for CO_2, and 0.03 for the mixture of gases that is terrestrial air (see for references Hansen & Travis 1974, Stam, Aben & Helderman 2002). The depolarization factor is slightly wavelength dependent (for wavelength dependent values, see Bates (1984)).

Both ρ and n_{i} are only slightly wavelength dependent, and the wavelength dependence of $b^{\text{sca}}_{\text{m}}$ can therefore be approximated as $1/\lambda^4$. At short wavelengths (ultraviolet, blue), incident sunlight is thus scattered very efficiently by atmospheric gases, while at long wavelengths (red) the incident sunlight can penetrate deep into the atmosphere without being scattered. For example, at 0.4 μm, the molecular scattering optical thickness of the Earth's atmosphere is about 0.4, while at 0.8 μm, it is about 0.02. The wavelength dependence of $b^{\text{sca}}_{\text{m}}$ explains the colour of the Earth's cloudfree day sky: the blue part of the incident sunlight is scattered much more efficiently at all altitudes than the red part. An observer at the surface will see the most of the blue sunlight coming from all directions, while most of the red sunlight is still contained in the direct beam (and is thus observed when looking directly at the Sun). Not all planetary atmospheres are blue, because the colour of the most efficiently scattered sunlight also depends on the molecular absorption optical thickness of the atmosphere, and on the aerosol scattering and absorption optical thicknesses. On Mars, for example, the day sky is not blue, but more reddish, because the molecular scattering optical thickness of the atmosphere is very small (Mars's surface pressure is about 1% of that on Earth, and b_m^{sca} of the atmosphere is thus only about 0.004 at 0.4 μm), while the scattering optical thickness of the dust aerosol is relatively high (about 0.1 in the visible, and increasing towards red wavelengths).

The molecular absorption optical thickness of an atmospheric layer can be calculated using

$$b_m^{abs}(\lambda) = \sigma_m^{abs}(\lambda)n_m = \frac{N_{av}\, \sigma_m^{abs}(\lambda)}{R} \int_{\Delta z} \eta(z)\frac{p(z)}{T(z)}dz, \qquad (12)$$

with η the volume mixing ratio of the absorbing gas, i.e. the fraction of the amount of absorbing gas molecules to the total amount of gas molecules. For example, on Earth, η of a well-mixed (i.e. altitude independent) gas like oxygen (O_2) is 0.21. The molecular absorption cross-section, σ_m^{abs}, usually depends not only on λ, but also on the ambient pressure and temperature. In Eq. (12), we simply assume that σ_m^{abs} is constant across Δz. Molecular absorption cross-sections usually show rapid variations with wavelength across limited wavelength regions. Such regions with absorption features are called absorption bands. As an example, Figure 3 shows the molecular absorption optical thickness b_m^{abs} of the Earth's atmosphere, across the so-called O_2 A-band (other, less strong, absorption bands of O_2 are referred to as the B and the γ-band, see also Figure 5), together with the atmospheric molecular scattering optical thickness b_m^{sca}. In the strongest absorption lines, b_m^{abs} is on the order of 300, and at these wavelengths, virtually no sunlight will emerge from the top or the bottom of the atmosphere. However, because the spectral resolution of most instruments that are used for planetary observations is much lower than the width of individual spectral lines, a spectral pixel will still receive light at wavelengths where b_m^{abs} is relatively small (see Figures 5 and 6).

The column number density of the atmospheric aerosol particles, n_a, is basically a free parameter: it varies from planet to planet, and in each planetary atmosphere it usually varies from location to location, and from time to time. The scattering and absorption cross-sections of aerosol particles, b_a^{sca} and b_a^{abs} respectively, depend on their sizes, composition, and shape. The cross-sections can be calculated, for example, using Mie-theory (van de Hulst 1957; de Rooij & van der Stap 1984) when the aerosol particles can be assumed to be homogeneous spheres or to consist of homogeneous, concentric shells, using the so-called T-matrix method for spheroidally shaped (oblate and prolate) particles (including spheres) (Mishchenko, Travis & Mackowski 1996), using the Discrete Dipole Approximation (DDA) method for irregularly shaped particles (Draine 2000, Draine & Flatau 1994) (because of computation time limitations, this method is as yet only used for relatively small particles), using geometrical optics (also called ray-tracing) for particles that are large with respect to the wavelength, or using laboratory measurements (see Volten et al. 2005).

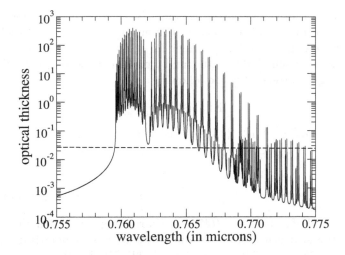

Figure 3. *The molecular absorption optical thickness, b_m^{abs}, across the O_2 A-band (solid line), together with the molecular scattering optical thickness, b_m^{sca} (dashed line), as calculated for a model Earth atmosphere.*

2.4.2 Single scattering albedo

The single scattering albedo ω of the particles in an atmospheric layer is given by

$$\omega(\lambda) = \frac{\sigma^{sca}(\lambda)}{\sigma^{abs}(\lambda) + \sigma^{sca}(\lambda)} = \frac{\sigma^{sca}(\lambda)}{\sigma^{ext}(\lambda)} = \frac{b^{sca}(\lambda)}{b^{ext}(\lambda)} = \frac{b^{sca}(\lambda)}{b^{abs}(\lambda) + b^{sca}(\lambda)} \quad (13)$$

A fraction ω of the flux that is intercepted by the particles is thus scattered, while the fraction $1 - \omega$ is absorbed. Assuming the atmospheric layer is in thermal equilibrium, the absorbed energy is emitted again (otherwise, the layer would heat up), but usually at different (longer) wavelengths.

2.4.3 Single scattering matrix

The single scattering matrix describes the angular distribution of the light that is intercepted by the particles in the atmospheric layer, and that they scatter around. The single scattering matrix of molecules, spherically symmetric particles, or irregularly shaped particles (with equal amounts of particles and their mirror particles) that are randomly oriented in space has the following shape (Hovenier, van der Mee & Domke 2004, Hansen & Travis 1974)

$$\mathbf{M}(\lambda, \Theta) = \begin{bmatrix} a_1(\lambda, \Theta) & b_1(\lambda, \Theta) & 0 & 0 \\ b_1(\lambda, \Theta) & a_2(\lambda, \Theta) & 0 & 0 \\ 0 & 0 & a_3(\lambda, \Theta) & b_2(\lambda, \Theta) \\ 0 & 0 & -b_2(\lambda, \Theta) & a_4(\lambda, \Theta) \end{bmatrix}. \quad (14)$$

Here, Θ is the single scattering angle (with $\Theta = 0°$ denoting scattering in the forward direction). Each scattering matrix is normalized such that the average of the phase function (element M_{11} or a_1) over all directions is unity. The matrix describing light scattering by gaseous molecules (usually referred to as Rayleigh scattering) is the following

$$\mathbf{M}_\mathrm{m}(\Theta) = \frac{3(45+\epsilon)}{180+40\epsilon}$$

$$\begin{bmatrix} (1+\cos^2\Theta) + 12\epsilon/(45+\epsilon) & \cos^2\Theta - 1 & 0 & 0 \\ \cos^2\Theta - 1 & 1 + \cos^2\Theta & 0 & 0 \\ 0 & 0 & 2\cos\Theta & 0 \\ 0 & 0 & 0 & E\cos\Theta \end{bmatrix}, \quad (15)$$

where E stands for $10(9 - \epsilon)/(45 + \epsilon)$. Parameter ϵ is called the anisotropy factor of the scattering molecule and is related to depolarization factor ρ through $\epsilon = 45\rho/(6 - 7\rho)$ (Stam et al. 2002). For brevity, we have omitted the λ-dependence of $\mathbf{M}_\mathrm{m}$ and ϵ in Eq. (15) (ϵ and hence the elements of $\mathbf{M}_\mathrm{m}$ depend only slightly on λ across the visible). The flux (phase function) and degree of polarization P of unpolarized light with $\lambda = 0.55$ μm that is singly scattered by gaseous molecules of terrestrial air, is shown in Figure 4. For molecules, both functions are symmetric around $\Theta = 90°$. At $\Theta = 0°$ and $180°$, $P = 0$ because of symmetry. At $\Theta = 90°$, P reaches almost 95% (not 100% because of the non-zero depolarization factor). The single scattering matrix of aerosol particles, $\mathbf{M}_\mathrm{a}$ can be obtained using the methods mentioned before (Mie-theory, T-matrix method, DDA method, geometrical optics, or laboratory measurements). As examples, Figure 4 shows the flux and degree of polarization of light singly scattered by small water cloud droplets at three wavelengths. The cloud droplets are distributed in size according to the standard size distribution of Hansen & Travis (1974), with an effective radius of 2.0 μm, and an effective variance of 0.1. The figure has been taken from Stam (2008). Again, at $\Theta = 0°$ and $180°$, $P = 0$ because of symmetry. The droplets show a strong forward ($\Theta = 0°$) scattering phase function, which is characteristic for particles that are large with respect to the wavelength of the incident light. The angular features in the phase function around $\Theta = 140°$ are usually referred to as primary rainbows. These rainbows also show up, even more strongly, in P. Furthermore, P can be seen to change sign, indicative for a change in the direction of polarization, at several scattering angles.

The scattering matrix of a mixture of different types of particles, for example gaseous molecules and aerosol particles (as will often be encountered in an atmospheric layer), can be calculated using appropriate weights, as follows

$$\mathbf{M}(\lambda, \Theta) = \frac{b_\mathrm{m}^\mathrm{sca}(\lambda)\mathbf{M}_\mathrm{m}(\lambda, \Theta) + b_\mathrm{a}^\mathrm{sca}(\lambda)\mathbf{M}_\mathrm{a}(\lambda, \Theta)}{b_\mathrm{m}^\mathrm{sca}(\lambda) + b_\mathrm{a}^\mathrm{sca}(\lambda)}. \quad (16)$$

In case an atmospheric layer contains more than two types of particles, Eq. (16) can be straightforwardly extended.

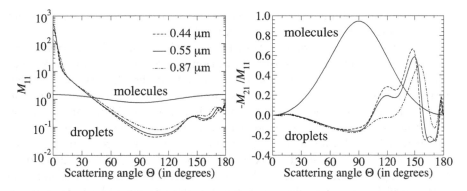

Figure 4. *On the left: the phase function (element $M_{11} = a_1$) and on the right: the degree of polarization ($-M_{12}/M_{11} = -b_1/a_1$) of unpolarized incident light that is singly scattered by gaseous molecules and water cloud droplets, at $\lambda = 0.44\ \mu m$ (droplets only), 0.55 μm, and 0.87 μm (droplets only).*

2.4.4   Surface reflection

The reflection properties of the surface, finally, are fully described by a wavelength dependent reflection matrix and surface albedo. Usually the surface is assumed to reflect in a Lambertian manner, i.e. isotropically and completely depolarizing. The Lambertian reflection matrix is given by a matrix of which all elements equal zero, except the (1,1)-element, which equals the wavelength dependent albedo, A_S. Some more realistic modelling for surface reflection matrices has been done, in particular to describe anisotropic reflection (usually referred to as the bidirectional reflection function, or the BRDF), yet very little modelling has been done on the polarizing properties of surfaces (also because of the lack of observations of these properties). An example of a type of surface for which a full reflection matrix is available, is a Fresnel (specular) reflecting surface (for formulae, see e.g. Deuze, Herman & Santer 1989).

2.5   Radiative transfer algorithms

Knowing b^{ext}, ω, and **M** of each atmospheric layer, and knowing the reflection properties (matrix and albedo) of the surface (if there is any), one can use standard radiative transfer algorithms to calculate the planetary scattering matrix **S**, and hence the flux vector of the starlight that is reflected by the planet. With these methods, the spherical shape of the planet is usually ignored. Instead, the reflection matrix of a locally plane-parallel atmosphere possibly with an underlying surface is calculated, for local solar zenith angle ranging from 0° (for locations with the star overhead) to 90° (for locations where the star is either rising or setting), for local viewing angles ranging from 0° (the center of the disk) to 90° (the edge of the planetary disk), and for the various azimuthal angles. The local reflection matrices can be inte-

grated across the illuminated and visible part of the planetary disk to yield
the scattering matrix of the planet as a whole.

Note that in most radiative transfer algorithms, radiation is treated as
a scalar: the radiation's state of polarization is ignored, and only fluxes are
calculated. The advantage of this treatment is that the algorithms are much
simpler and require significantly less computing time. This might seem like
a good idea in cases where no polarimetry is performed. However, as can be
seen from the shape of the single scattering matrices, incident unpolarized
light gets polarized upon a single scattering. When this singly scattered light
is scattered a second time, the flux of the twice scattered light will depend on
the state of polarization of the singly scattered light! Indeed, if light is multiple
scattered within an atmosphere, ignoring the polarization will lead to errors in
calculated fluxes (Mishchenko, Lacis & Travis 1994, Stam & Hovenier 2005).

We use an efficient adding-doubling radiative transfer algorithm (de Haan,
Bosma & Hovenier 1987) that fully includes polarization and multiple scattering, to calculate the reflection matrix of a locally plane-parallel planet. To
integrate the local results efficiently across the planetary disk, we use a special disk integration method that treats the planet as if it were a scattering
particle (Stam et al. 2006). With this method, we calculate an expansion of
the scattering matrix elements into generalized spherical functions (see, e.g.
Hovenier et al. 2004). From these (phase angle independent) coefficients, the
planetary scattering matrix **S** can rapidly be obtained at arbitrary planetary
phase angles. The disadvantage of our algorithm is that it can handle hori-
zontally homogeneous planets only. By combining results for different homogeneous planets, we can simulate quasi-inhomogeneous planets, for example
a cloudfree planet covered with 40% vegetation and 60% ocean (for sample
simulations, see Stam 2008).

3 Flux and polarization spectra

3.1 Observations of the Earth

Before focussing on results of numerical simulations of gaseous and terrestrial
type exoplanets, we'll present observations of sunlight that is reflected by
the Earth, as measured by an Earth remote-sensing instrument. Although
these flux and polarization observations pertain only to a small part of the
planet, and can therefore not directly be compared with (future) exoplanet
observations, they clearly show the spectral structures that can be expected
for spectra of (terrestrial) exoplanets. We choose these Earth spectra, and not
spectra of e.g. a gaseous planet, because as far as we know, for none of the other
solar-system planets are there high-resolution spectra of both the flux and the
polarization available. The observed flux and polarization spectra are shown
in Figure 5. The flux spectra are measured by the Global Ozone Monitoring
Experiment (GOME) onboard the European Space Agency's (ESA) second

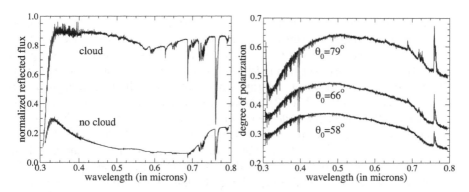

Figure 5. *On the left: the reflected flux of two regions on Earth. The upper curve was measured above a region that was completely covered with clouds, and the lower curve was measured above a cloud-free region of the Netherlands that is covered with vegetation. In both cases, the solar zenith angle θ_0 is about 34°, and the observed light is coming from the nadir direction. On the right: polarization P of the cloud free zenith sky for three values of θ_0.*

Earth Remote-sensing Satellite (ERS-2) satellite (launched in 1995) looking down at the Earth from space (in July 1995). The observed regions have a size of about 40 × 80 km². The fluxes have been normalized to the incoming solar flux (that was also measured by GOME). The polarization spectra are measured by a copy of GOME looking up to the zenith sky from the Earth's surface (in the Netherlands) in April, 1997 (Aben et al. 1999). The spectral resolution of GOME is 0.0002 to 0.0004 μm.

Each flux and polarization spectrum can be thought of as consisting of a continuum with superimposed high-spectral resolution features. The flux in the continuum of the 'cloudy' spectrum is high because the clouds reflect most of the incident sunlight. The low flux below about 0.34 μm is mostly due to aborption by ozone (O_3). This absorption band is called the Huggins band. The shallow depression in the continuum flux between $\lambda = 0.5$ and 0.7 μm is the Chappuis absorption band of ozone. In the 'no cloud' flux spectrum, the continuum at the shortest wavelengths is determined by the Huggins band, while with increasing wavelength above about 0.34 μm, the continuum flux steadily decreases, until $\lambda \approx 0.5$ μm. This decrease is entirely due to the decrease of b_m^{sca} with wavelength (see Eq. (11)). The bump in the spectrum around $\lambda = 0.54$ μm (green light) is due to reflection by vegetation, just like the steep increase in the near infrared ($\lambda > 0.7$ μm). The latter increase is usually referred to as the 'red edge' and is assumed to be representative for terrestrial-type vegetation (for a thorough introduction on the red edge, see Seager et al. (2005);Arnold (2008)). On planets around stars with different spectra than the Sun, the albedo of vegetation might differ from that found on Earth (Tinetti, Rashby & Yung 2006);(Kiang et al. 2007).

The shape of the continuum degree of polarization is strongly determined by the ratio of singly scattered to multiple scattered light (see also Aben et al. 1999): multiple scattering tends to lower P. At the shortest wavelengths, in the Huggins absorption band, b_m^{abs} is huge, and the feeble observed light will be mostly singly scattered light, with a relatively high degree of polarization. With increasing λ, b_m^{abs} decreases, and multiple scattering increases, lowering P. The subsequent decrease of b_m^{sca} with λ, however, leads to a decrease of multiple scattering, and hence an increase of P, for $\lambda > 0.32$ μm. At the longer wavelengths ($\lambda > 0.45$ μm), P is increasingly influenced by light scattering by aerosol particles (decreasing P in these cases), and at the longest wavelengths ($\lambda > 0.7$ μm), the 'red edge' signature of the vegetation is recognized in P. Albedo features of the surface show up in P of the zenith sky through light that has first been reflected upwards by the surface, and that is scattered downwards again within the atmosphere. Both the multiple reflections/scatterings and the usually only weakly polarizing influence of the surface lead to a decrease of P. In Section 3.2, we'll show the effect of the surface albedo on P of the starlight that is reflected to space using numerical simulations. The height of the continuum polarization in Figure 5 increases with the solar zenith angle θ_0, because with the zenith viewing direction, θ_0 equals the scattering angle of the singly scattered light, Θ. As can be seen in Figure 4, the degree of polarization of singly scattered light varies strongly with Θ, and for light singly scattered by gaseous molecules, the degree of polarization increases with decreasing $|\Theta - 90°|$. This explains the relation between P and θ_0 in Figure 5.

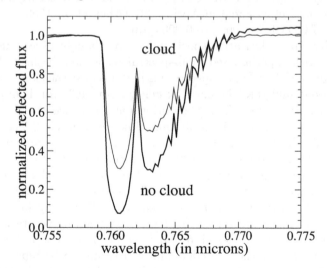

Figure 6. *The flux spectra of Figure 5 across the O_2 A-band, normalized at 0.758 μm. The thin line corresponds to the cloudy spectrum and the thick line to the cloud free spectrum.*

There are two types of high-spectral resolution feature in the flux and polarization spectra. The first type is most obvious above about 0.4 μm: dips in the flux spectra and matching peaks in the polarization spectra. These features are due to absorption by atmospheric gases: H_2O (e.g. around 0.60 μm, 0.65 μm, 0.73 μm, and 0.79 μm), and O_2 (the γ-band near 0.63 μm, the B-band around 0.68 μm, and the A-band around 0.76 μm). The origin of the peaks in the polarization lies in the decrease of multiple scattering when b_m^{abs} increases, as explained above for the O_3 Huggins absorption band. The behaviour of P across an absorption band is also influenced by the vertical distribution of atmospheric particles (as discussed by Stam et al. 1999): in the deepest part of an absorption band, the observed light has predominantly been scattered by particles in the upper atmospheric layers, and thus carries the polarization signatures of these high-altitude particles, while in the continuum, the observed light has mostly been scattered by particles in the lower atmospheric layers (because these are usually the optically thickest). Depending on the type of high-altitude particles, P could even be lower in the absorption band than in the continuum.

The shape and depth of absorption bands in the flux spectra is also sensitive to the composition and structure of the planetary atmosphere. This is clearly illustrated in Figure 6, where we show part of the flux spectra of Figure 5, normalized at 0.758 μm. In the figure, you can see how clouds decrease the depth of the O_2 A-band by efficiently reflecting sunlight back to space at high altitudes, and thus by reducing the effective b^{abs} of the atmosphere. In Earth observation, measurements of the depth of this absorption band are routinely used to derive cloud top altitudes, which is possible because the vertical distribution of O_2 is known (see e.g. Kuze & Chance 1994).

The second type of high-spectral resolution feature is found mostly below about 0.4 μm: peaks in the flux spectra and narrow dips in the polarization spectra. A careful look at these features reveals that they match the Fraunhofer lines in the solar spectrum (see Aben, Stam & Helderman 2001). That these lines show up in the flux spectra of scattered sunlight (that have been divided by the incoming solar spectrum) and the polarization spectra, is explained by rotational Raman scattering, an inelastic scattering process of gaseous molecules. Because the process is inelastic, the scattered light is distributed over a range (a few nm) of wavelengths, thus 'filling-in' the Fraunhofer lines in the scattered light. In Earth-remote sensing, this phenomenon is referred to as the Ring-effect, after one of the discoverers (Grainger & Ring 1962). Because the degree of polarization of the inelastically scattered light is lower than that of the elastically scattered light, the filled-up Fraunhofer lines have a lower degree of polarization than the continuum (for more details and references, see Stam et al. (2002); Aben et al. (1999). The fraction of rotational Raman scattered light depends on the type of molecule and on the wavelength; in the Earth's atmosphere, this fraction is about 0.04.

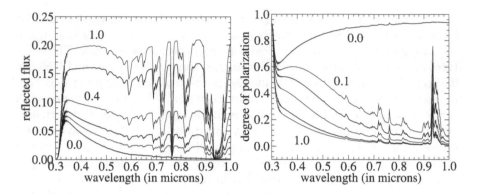

Figure 7. *The flux (left) and degree of polarization (right) of starlight reflected by cloud free Earth-like planets with Lambertian surfaces and wavelength independent surface albedos: 0.0, 0.1, 0.2, 0.4, 0.8, and 1.0. The planetary phase angle is 90°.*

3.2 Simulations of exoplanet spectra

Although flux and/or polarization spectra of gaseous exoplanets will most probably be available before those of Earth-like exoplanets, here we'll present numerical simulations of the latter first with the observations of the previous section fresh in our minds. In the following, we set the planetary radius r, the distance between the star and the Earth-like or gaseous planet D, and the distance between the planet and the observer d, all equal to one. In addition, we assume that $R^2\pi B_0=1$. According to Eq. (7), the stellar flux that is reflected by the planet thus equals $a_1/4$, which is the planet's geometric albedo A_G. The calculated reflected fluxes that we present here can easily be scaled using the parameters of any given planetary system. Our simulations do not include rotational Raman scattering, so there will be no high-spectral resolution peaks in the flux and dips in the polarization. The spectral resolution of our simulations is lower than that of the observations presented in Figure 5, namely 0.001 μm.

3.2.1 Earth-like exoplanets

Our model Earth-like exoplanets have pressure and temperature profiles according to a mid-latitude summer standard profile, and terrestrial-type air, with H_2O, O_3, and O_2 as absorbing gases (McClatchey et al. 1972). In Figure 7, we have plotted numerically simulated flux and polarization spectra of model planets with atmospheres without aerosol (or clouds) and Lambertian reflecting surfaces with wavelength independent albedos A_S ranging from 0.0 to 1.0 (details and more simulations can be found in Stam 2008). The planetary phase angle α is 90°. Thus half of the observable planetary disk is illuminated by the star. The probability of observing an exoplanet at or near this phase angle is relatively high. The shapes of the curves in Figure 7 are easy to understand with the knowledge from Section 3.1: we can identify the absorption bands of O_2, O_3, and H_2O (because the simulations run until $\lambda = 1.0$ μm, they include a strong H_2O-band between 0.9 and 1.0 μm, that cannot be seen in Figure 5). The H_2O and O_2-bands are relatively deep in Figure 7, because the atmosphere contains no clouds. Because most of the O_3 is in the Earth's stratosphere, above the clouds, the depth of O_3 absorption bands is less influenced by clouds.

The curves pertaining to the black surface ($A_S = 0.0$), show the effects of the atmosphere alone: the reflected flux is relatively high at the shortest wavelengths and decreases rapidly with increasing λ (approximately as $1/\lambda^4$). The degree of polarization P starts off high because of the absorption by O_3, then drops because of multiple scattering, and then increases with λ because of the decrease of b_m^{sca}, and hence the decrease of multiple scattering. At the longest wavelengths, b_m^{sca} is too small to have a significant contribution of multiple scattered light, and P reaches its single scattering value for $\Theta = 90°$ (see Figure 4). In Figure 7, the curves pertaining to the model planets with reflecting surfaces ($A_S > 0.0$) show a strong influence of the surface albedo, especially at longer wavelengths, where b_m^{sca} is small and the incident light can easily reach the surface and after that space again. In particular, because light reflected by our model surfaces is unpolarized, higher values of A_S add more unpolarized light to the reflected flux. To realistically model the influence of the surface reflection on the planetary flux and polarization signal, more realistic surface models, including surface polarization, have to be developed.

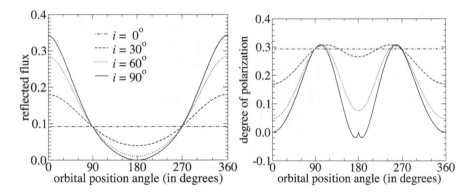

Figure 8. *The flux (left) and degree of polarization (right) of starlight with $\lambda = 0.44$ μm that is reflected by a cloud free Earth-like planet with $A_S = 0.4$ along the planetary orbit, for different orbital inclination angles i: $0°$ (face-on), $30°$, $60°$, and $90°$ (edge-on).*

Figure 8 shows the reflected flux and degree of polarization for $\lambda = 0.44$ μm of a cloud free Earth-like planet with $A_S = 0.4$ as functions of the planet's position in its orbit for different orbital inclinations angles i. When $i = 0°$, the orbit is seen face-on, thus everywhere along its orbit, the planet has a planetary phase angle α of 90° (resembling a 'half moon'), and yields the same flux and polarization signal (the absolute flux might of course change when the orbit is eccentric rather than circular). When $i = 90°$, the orbit is seen edge-on, and α varies between 0° and 180° along the planetary orbit. Obviously, with this orbital inclination angle, the flux and polarization variation along the orbit is maximum. Of course, the planet cannot be observed at those phase angles where it would be too close to its star. For example, at $\alpha = 0°$, the observer would see a fully illuminated planet, if the star were not in front of the planet, and at $\alpha = 180°$, the planet is in front of its star, with its nightside turned towards the observer (it is 'transiting' its star). The angular distance between the star and the planet is largest when $\alpha = 90°$, and when for every value of i, the maximum values of P occur. Because $\alpha = 90°$ at least twice every planetary orbit, independent of i, relatively high values of P can be observed at least twice every orbit.

In Figure 9, finally, we show the flux and degree of polarization of starlight that is reflected by Earth-like planets with vegetation or ocean covering their surfaces (see also Stam 2008). The ocean is black at all wavelengths, but has a Fresnel reflecting surface. For each of the two types of planet, we show curves for a clear and a completely cloudy atmosphere, for $\alpha = 90°$. The cloud has an optical thickness b_a^{ext} ($\approx b_a^{sca}$) of 10 at $\lambda = 0.55$ μm, and is located between 2 and 4 km of altitude. The phase function and polarization of light singly scattered by the cloud particles have been shown in Figure 4. For comparison, we have also plotted the $A_S = 0.0$ and $A_S = 1.0$ curves of Figure 7.

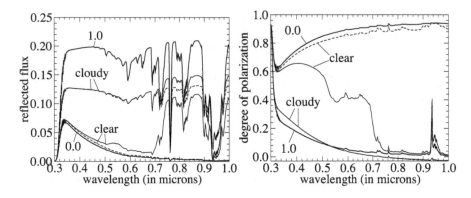

Figure 9. *Similar to Figure 7, except that the surfaces of the Earth-like model planets are covered with either vegetation or a Fresnel-reflecting ocean, and the atmospheres are either clear or completely cloudy. Also shown are the $A_S = 0.0$ and $A_S = 1.0$ curves of Figure 7. The planetary phase angle is 90°.*

The cloud free ocean planet is almost as dark as a black planet: only in the blue, the Fresnel reflection adds a little bit of flux to the total. This is somewhat surprising, since on images of the Earth taken from space, the reflection of the Sun can often be seen as a bright spot on the ocean (the 'sun-glint'). Our ocean planet might be this dark because our star is assumed to be at infinite distance from the planet ($D \gg R$). Thus you would not expect to find a mirror image of the star on the planet. It might also be that an ocean planet is indeed quite dark: averaged over the illuminated half of the planet, the 'star-glint' might increase the total flux only a little bit. In the polarization, the Fresnel reflection lowers P by more than 5% as compared to a black surface. The cloud free vegetation planet is easily recognized by the 'bump' and the 'edge'. Because the vegetation surface reflects unpolarized light, the edge strongly suppresses P at the longer wavelengths. The bump is also an obvious feature in P.

Adding clouds to a planet strongly increases the reflected flux, as can be seen in Figure 9. Even though the clouds are thick, the surface albedo still influences the total flux: the cloudy planets are much darker than a white ($A_S = 1.0$) planet, because the clouds diffusely transmit light towards the relatively dark surface, and the edge in the vegetation albedo can still be recognized. The clouds strongly decrease P (the two lines are indistinguishable), both because they increase the multiple scattering of light in the atmosphere, and because the degree of polarization of light singly scattered by the cloud particles is small and opposite in sign to that of the light scattered by the molecules (see Figure 4). This opposite direction of polarization also causes the change of polarization direction of the reflected starlight around 0.7 μm, which does not happen for the white planet.

3.2.2 Jupiter-like exoplanets

Our model gaseous exoplanets have Jupiter-like atmospheric pressure and temperature profiles, and a similar composition (mainly H_2, with 0.18% CH_4). The molecular scattering optical thickness of our model atmosphere is more than 20 at 0.4 μm, and decreases to 0.51 at 1.0 μm (see Eq. (11)). The atmosphere is bounded below by a totally absorbing layer. We will show results for three types of atmospheres: model 1 contains only gaseous molecules (the 'clear model'), model 2 is similar to model 1, except it also has a cloud layer in the troposphere (the 'cloud model'), and model 3 is similar to model 2, except it also has a thin haze in the stratosphere (the 'haze model'). For the cloud layer, $b_a^{ext} = b_a^{sca} = 6$ at $\lambda = 0.7$ μm, and for the haze, $b_a^{ext} = b_a^{sca} = 0.25$ at $\lambda = 0.7$ μm. Details on the simulations can be found in Stam et al. (2004).

The cloud and haze particles are homogeneous, spherical particles distributed in size according to the standard size distribution of Hansen & Travis (1974). For the cloud particles, $r_{eff} = 1.0$ μm and $v_{eff} = 0.1$. Their refractive index is 1.42, which is typical for NH_3 ice. For the haze particles, $r_{eff} = 0.5$ μm and $v_{eff} = 0.01$, and their refractive index is 1.66. We calculate the wavelength dependent extinction cross-sections, single scattering albedos, and scattering matrices of the cloud and haze particles using Mie-theory (de Rooij & van der Stap 1984). Figure 10 shows the phase functions (matrix element M_{11}) and degree of (linear) polarization ($-M_{21}/M_{11}$) of light singly scattered by the gaseous molecules, the cloud particles and the haze particles at $\lambda = 0.7\ \mu$m.

In Figure 11, we show the flux and degree of polarization of starlight reflected by the three model planets (see also Stam et al. 2004). The absorption bands in the curves are all due to methane (CH_4). These bands do not show up in the spectra of the Earth-like planets, because the absorption cross-section σ_m^{abs} across these bands is very small and the total amount of CH_4, n_m, in the Earth's atmosphere is too small to have a noticeable value of b_m^{abs}. See Eq. (3). In a Jupiter-like atmosphere, σ_m^{abs} is still very small, but n_m of CH_4 is huge, resulting in significant values of b_m^{abs}.

The continuum flux reflected by the gaseous atmosphere (model 1) decreases smoothly with wavelength, because of the decrease of b_m^{sca}. Adding the cloud layer to the atmosphere (model 2), makes little difference in the flux at the short wavelengths, because there the flux is mainly determined by b_m^{sca} of the atmospheric layers above the clouds: the cloud is 'too deep into the atmosphere' to show up at these wavelengths. With increasing λ and, hence, decreasing b_m^{sca}, the contribution of the cloud to the flux increases. At the longest wavelengths, the cloud layer doubles the reflected flux in the continuum. Adding the haze layer to the atmosphere (model 3) increases the reflected flux somewhat, mostly in the absorption bands. The effects of the haze on the flux are small because the haze is very thin.

The continuum polarization of our model 1 planet starts off at about 32 % at $\lambda = 0.4\ \mu$m, despite the large amount of multiple scattered light at these wavelengths. This relatively high degree of polarization is basically due to the contribution of the singly scattered light, assuming that most of the multiple scattered light will be unpolarized because of the multiple scatterings. With increasing λ, the continuum P increases, because b_m^{sca}, and hence the multiple scattering, decreases. In the continuum at the longest wavelengths, P does not reach its single scattering value (see Figure 10), because b_m^{sca} is still large enough (about 0.5) for multiple scattering to occur. The single scattering value for P is reached only in the deepest absorption bands.

As with the flux, adding the cloud layer (model 2) does not influence the continuum P at the shortest wavelengths, because the cloud is located too deep in the atmosphere. Only at $\lambda \approx 0.5\ \mu$m, does the cloud start to suppress P, because of the multiple scattering within the cloud, and because of the low degree of polarization of the light singly scattered by the cloud particles (see Figure 10). In the deepest absorption band (around $\lambda = 0.9\ \mu$m), the incident starlight cannot reach the cloud, and P behaves as if the atmosphere were cloud free. The haze (model 3), which hardly left a trace in the reflected flux, shows up clearly in P, and, because of its high altitude, even at the shortest wavelengths. The influence of the haze on P is especially apparent in the absorption bands: these are very shallow, mainly because the light that is reflected at these wavelengths carries the single scattering signature of the particles in the highest atmospheric layers, and in model 3 these are the haze

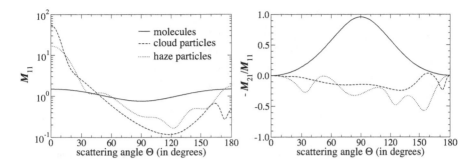

Figure 10. *Similar to Figure 4, except for 'air' molecules (H$_2$), and Jupiter-like cloud droplets and haze particles, at $\lambda = 0.7$ μm.*

particles. See Figure 10.

Interestingly, while the haze is virtually invisible in the reflected flux, it leaves clear traces in the polarization. This is even more obvious from Figure 12, where we have plotted the flux and degree of polarization averaged between $\lambda = 0.65$ and 0.95 μm, as functions of the planetary phase angle α. The curves for the different planets show that whereas the fluxes of model 2 and model 3 are almost indistinguishable from each other at every phase angle (except at the very small phase angles, where the planet cannot be observed because there it is too close to its star), the polarization curves are very different from each other. Both polarization curves are also very different from the curve of model 1, which would help identifying a clear planetary atmosphere: even though in theory model 1's flux curve stands out from those of models 2 and 3, in practice it will be very difficult to derive fluxes so accurately, not in the least because many parameters that determine the flux, such as planet size, will be uncertain. Figures 11 and 12 strengthen the idea that the combination of flux with polarization observations provides the best characterization of exoplanetary atmospheres from reflected starlight.

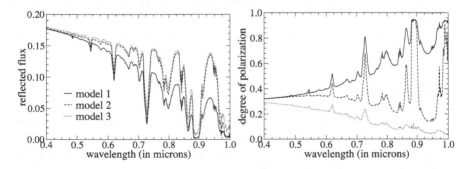

Figure 11. *Similar to Figure 9, except for the three Jupiter-like model planets.*

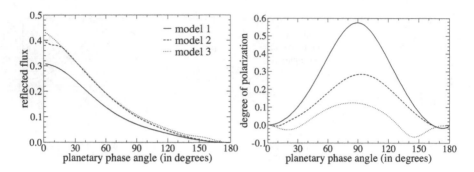

Figure 12. *The flux (left) and degree of polarization (right) of the three model atmospheres as functions of the planetary phase angle, averaged over the wavelength region between 0.65 and 0.95 μm.*

4 Summary

In the coming years, we will see more and more direct detections of radiation from exoplanets. Because with direct detections we can discover exoplanets that are too small or too far away from their star to be detected with the currently used indirect methods, they will open up an as-yet uncharted part of the planetary parameter space. Direct detections will help to answer the question whether our solar system, with its small, terrestrial inner planets and large, gaseous outer planets is special or rather common. By analyzing flux and/or polarization spectra of the planetary radiation, we can retrieve physical parameters of exoplanets, and study the composition and structure of their atmospheres and surfaces (for terrestrial planets). For this analysis and for the optimization of the instruments, numerical simulations of exoplanet signals are essential. The spectra of Earth-like and Jupiter-like exoplanets that were shown and discussed in this chapter are only starting points, because we do not know what other planets in other planetary systems look like. The radiative transfer processes will however be the same, and will yield similar spectral features to the ones described here. Indeed, direct detections will widen the field of comparative planetology, not only teaching us about exoplanets, but also about the planets around our own Sun.

References

Aben, I., Stam, D. & Helderman, F. (2001), 'The ring effect in skylight polarization', *Geophys. Res. Lett.* **28**, 519–522.

Aben, I. et al. (1999), 'Spectral fine-structure in the polarization of skylight', *Geophys. Res. Lett.* **26**, 591–594.

Arnold, L. (2008), 'Earthshine observation of vegetation and implication for life detection on other planets. A Review of 2001-2006 Works', *Space Sci.*

Reviews **135**, 323–333.

Bates, D. (1984), 'Rayleigh scattering by air', *Planet. Space Sci.* **32**, 785–790.

Beuzit, J.-L. et al. (2006), 'SPHERE: A 'Planet Finder' instrument for the VLT', *The Messenger* **125**, 29–+.

Burrows, A., Sudarsky, D. & Lunine, J. (2003), 'Beyond the T Dwarfs: Theoretical spectra, colors, and detectability of the coolest brown dwarfs', *Astrophys. J.* **596**, 587–596.

Charbonneau, D. et al. (2000), 'Detection of planetary transits across a sun-like Star', *Astrophys. J. Lett.* **529**, L45–L48.

Charbonneau, D. et al. (2002), 'Detection of an extra-solar planet atmosphere', *Astrophys. J.* **568**, 377–384.

Chauvin, G. et al. (2004), 'A giant planet candidate near a young brown dwarf. Direct VLT/NACO observations using IR wavefront sensing', *Astron. Astrophys.* **425**, L29–L32.

de Haan, J., Bosma, P. & Hovenier, J. (1987), 'The adding method for multiple scattering calculations of polarized light', *Astron. Astrophys.* **183**, 371–391.

de Rooij, W. & van der Stap, C. (1984), 'Expansion of Mie scattering matrices in generalized spherical functions', *Astron. Astrophys.* **131**, 237–248.

Deuze, J., Herman, M. & Santer, R. (1989), 'Fourier series expansion of the transfer equation in the atmosphere-ocean system', *J. Quant. Spectrosc. Radiat. Transfer* **41**, 483–494.

Draine, B. (2000), 'The discrete dipole approximation for light scattering by irregular targets', *Light Scattering by Nonspherical Particles: Halifax Contributions* pp. 131.

Draine, B. & Flatau, P. (1994), 'Discrete-dipole approximation for scattering calculations', *J. Optical Soc. Am. A* **11**, 1491–1499.

Grainger, J. & Ring, J. (1962), 'Anomalous Fraunhofer line profiles', *Nature* **193**, 762–+.

Hansen, J. & Hovenier, J. (1974), 'Interpretation of the polarization of venus', *J. Atmos. Sci.* **31**, 1137–1160.

Hansen, J. & Travis, L. (1974), 'Light scattering in planetary atmospheres', *Space Sci. Rev.* **16**, 527–610.

Hovenier, J. & van der Mee, C. (1983), 'Fundamental relationships relevant to the transfer of polarized light in a scattering atmosphere', *Astron. Astrophys.* **128**, 1–16.

Hovenier, J. W., van der Mee, C. & Domke, H. (2004), *Transfer of polarized light in planetary atmospheres; basic concepts and practical methods*, Kluwer, Dordrecht; Springer, Berlin.

Kemp, J. et al. (1987), 'The optical polarization of the sun measured at a sensitivity of parts in ten million', *Nature* **326**, 270–273.

Kiang, N. et al. (2007), 'Spectral signatures of photosynthesis. II. coevolution with other stars and the atmosphere on extra-solar worlds', *Astrobiology* **7**, 252–274.

Knutson, H. et al. (2007), 'A map of the day-night contrast of the extra-solar planet HD 189733b', *Nature* **447**, 183–186.

Kuze, A. & Chance, K. (1994), 'Analysis of cloud top height and cloud coverage from satellites using the O2 A and B bands', *J. Geophys. Res.* **99**, 14481–+.

Lafreniére, D., Jayawardhana, R. & van Kerkwijk, M. H. (2008), 'Direct imaging and spectroscopy of a planetary mass candidate companion to a young solar analog', *Submitted to Astrophys. J. Lett.* pp. L–L.

Mayor, M. & Queloz, D. (1995), 'A Jupiter-mass companion to a solar-type star', *Nature* **378**, 355–+.

McClatchey, R. et al. (1972), *Optical Properties of the Atmosphere, AFCRL-72.0497*, U.S. Air Force Cambridge Research Labs.

Mishchenko, M., Lacis, A. & Travis, L. (1994), 'Errors induced by the neglect of polarization in radiance calculations for Rayleigh-scattering atmospheres', *J. Quant. Spectrosc. Radiat. Transfer* **51**, 491–510.

Mishchenko, M., Travis, L. & Mackowski, D. (1996), 'T-matrix computations of light scattering by nonspherical particles: a review', *Journal of Quantitative Spectroscopy and Radiative Transfer* **55**, 535–575.

Seager, S., Whitney, B. & Sasselov, D. (2000), 'Photometric light Curves and polarization of close-in extra-solar giant planets', *Astrophys. J.* **540**, 504–520.

Seager, S. et al. (2005), 'Vegetation's red edge: A possible spectroscopic biosignature of extraterrestrial plants', *Astrobiology* **5**, 372–390.

Stam, D. (2008), 'Spectropolarimetric signatures of Earth-like extra-solar planets', *Astron. Astrophys.* **482**, 989–1007.

Stam, D., Aben, I. & Helderman, F. (2002), 'Skylight polarization spectra: Numerical simulation of the Ring effect', *J. Geophys. Res. (Atmospheres)* **107**, 4419–+.

Stam, D. & Hovenier, J. (2005) , 'Errors in calculated planetary phase functions and albedos due to neglecting polarization', *Astron. Astrophys.* **444**, 275–286.

Stam, D., Hovenier, J. & Waters, L. (2004), 'Using polarimetry to detect and characterize Jupiter-like extra-solar planets', *Astron. Astrophys.* **428**, 663–672.

Stam, D. et al. (1999), 'Degree of linear polarization of light emerging from the cloudless atmosphere in the oxygen A band', *J. Geophys. Res.* **104**(13), 16843–16858.

Stam, D. et al. (2006), 'Integrating polarized light over a planetary disk applied to starlight reflected by extra-solar planets', *Astron. Astrophys.* **452**, 669–683.

Swain, M., Vasisht, G. & Tinetti, G. (2008), 'The presence of methane in the atmosphere of an extra-solar planet', *Nature* **452**, 329–331.

Tinetti, G., Rashby, S. & Yung, Y. (2006), 'Detectability of red-edge-shifted vegetation on terrestrial planets orbiting M stars', *Astrophys. J. Lett.* **644**, L129–L132.

van de Hulst, H. C. (1957), *Light Scattering by Small Particles*, J. Wiley and

Sons, New York.

Volten, H. et al. (2005), 'WWW scattering matrix database for small mineral particles at 441.6 and 632.8nm', *J. Quant. Spectrosc. Radiat. Transfer* **90**, 191–206.

Part II

Formation and Evolution of Planetary Systems

Proto-planetary discs – current problems and directions

J S Greaves

SUPA, University of St Andrews, School of Physics & Astronomy, North Haugh, St Andrews, KY16 9SS, United Kingdom
STFC Advanced Fellow

1 Introduction

Discs are a natural part of the star formation process – the collapse of a rotating cloud core plus the conservation of angular momentum imply that material can not fall directly onto a growing protostar. Instead, material drifts down from the protostellar envelope to a smaller flattened disc of orbiting gas and dust particles, and this material then accretes, building up the young star towards its final mass somewhere in the pre-main sequence stage. Residual material in the disc is the reservoir for planet formation, and the disc origin is confirmed by finding the Sun's planets orbiting in a plane. Discovering extra-solar discs has taken surprisingly long. In the 1960's and 1970's, discs were inferred from infrared emission around young stars (i.e. from orbiting dust grains re-emitting absorbed stellar photons) and from scattering patterns of polarized light. The first true discovery was made with the coronographic imaging of the thin edge-on disc of beta Pictoris (Smith & Terrile 1984). At an age of about 10 Myr, we see the characteristic flattened disc plane, but on a much larger scale than in the solar system – particles are spread out to hundreds of AU from the star (possibly as a remnant of the protostellar envelope epoch), compared to the Kuiper Belt of cometary planetesimals in a solar system, the bulk of which lies within 50 AU. More overview and historical aspects may be found in Greaves (2005) and references therein.

Subsequently, many circumstellar discs have been imaged in nearby star formation regions, ranging from the diffuse Taurus association up to the rich Orion Nebula Cluster. In the optical, discs are revealed by scattering of stellar light, or in silhouette when they block the light of e.g. a bright background nebula. From the near-infrared out to the radio the thermal emission of dust grains is detected, ranging from hot dust (close to sublimation) near the star out to cold particles at tens to hundreds of AU. Thus choice of wavelength dictates the disc properties measured to a large extent, and this is also true of optical scattering, which can be seen far from the star in favourable scattering orientations even when the grains are sparse. The longest wavelengths are the best guide to disc mass, since the thermal emission is optically thin from the far-infrared longwards (Wood et al. 2002) – however, the most detailed images are in the optical and near-IR (or with radio interferometry), because of the small apparent size of circumstellar discs and better diffraction limit.

Many deductions about discs have been made from spatially unresolved data, i.e. plotting and fitting the spectral energy distribution, or SED, of the emission from the star-disc system. A stellar photosphere is a good approximation to a blackbody typically peaking shortwards of one micron, while the circumstellar material has a broad spectrum with a peak somewhere in the infrared, depending on the grain temperatures and hence orbital distances. A simple classification scheme uses the shift in the dominant emission, progressing from long-wavelength Class 0 protostellar objects (gas envelope, embedded disc and new protostar) to Class I (remnant envelope, bright disc, growing protostar) to Class II (disc and pre-main-sequence star) ending with Class III (remnant disc, optical star contracting on to main sequence). During this evolution, dust is coagulating in the disc from sub-micron interstellar medium (ISM) sizes to microns and millimetres, and sedimenting down towards the disc mid-plane. There must be a balance of such growth, and the loss of grains by light pressure and drag forces and destruction when the fragile conglomerates collide, but with a net upwards size-trend to produce planetesimals, out of which planets form (Ford 2010, Chapter 7).

Observations can also be made of the gas, which at early times is the dominant component of the disc, inheriting the gas-to-dust mass ratio of order 100 in the ISM. The gas is mainly molecular hydrogen in the cold dark disc mid-plane, with trace molecules such as carbon monoxide (CO), hydrogen cyanide (HCN) etc. made from the common universal elements. Hotter layers exist at the top and bottom of the disc where starlight penetrates, and where UV or X-rays from the active young star split the molecules, the resulting energetic atoms tend to evaporate. Thus in time gas is dispersed from the disc, either internally by the star or because of external irradiation from other stars in the cluster. The prime culprit is stellar UV or X-ray light, particularly associated with young stars with surface activity, strong coronae etc. Although gas observations are consequently hard to interpret in terms of disc mass, with many atomic and molecular excitation processes occurring, the spectra may be resolved in frequency and so more simply interpreted to show where the gas is located. A characteristic double-peaked pattern like a tooth is seen for gas on Keplerian orbits in a moderately flat disc, owing to a pile-up of line-of-sight velocities that occur both from the slow outer disc rotation and projections of faster rotation speeds further in. From such spectral profiles, the

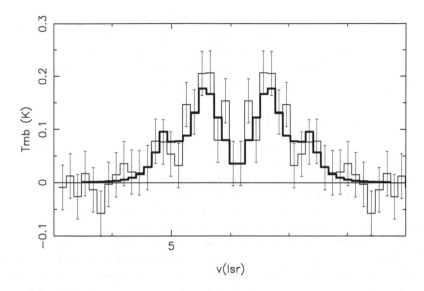

Figure 1. *Spectrum of the J=3-2 rotational transition of the HCO⁺ molecule in the disc of DM Tau (Greaves 2004). The thin-line histogram is the observed data (folded about the line centre to increase the signal-to-noise) and the thick-line histogram is from a model disc; the best-fit outer radius is ≈ 140 − 300 AU and the inclination is ≈ 32 − 45°.*

disc inclination, outer radius and presence of any inner hole may be deduced, without spatially resolving the structure (Figure 1).

2 Disc properties

Having measured disc properties through imaging, dust SEDs and gas spectra, interesting questions related to planet formation can be posed. These include: – are the discs suitable for planets to form? – are there signs of ongoing planet formation? – and, can we observe the planets as they grow? The first of these aspects has the most relevant data, and some basic properties have been established by simple methods. For example, the timescales of various phases can be found by counting the number of discs of each type within a young star cluster assumed to be of co-eval origin. Then by bootstrapping the counts to those for pre-main-sequence stars with 'proper' ages from isochrones (i.e. their contraction on Hayashi tracks down onto the main-sequence), the earlier eras can be quantified. Ballpark results are that Class 0 and I protostars last for a few 10^{4-5} years, and Class II and III stars co-exist over ages of around 1-10 million years. It is presumed that Class III objects have somehow lost most of their disc material (perhaps by radiation stripping or destructive perturbations caused by other cluster stars) and so their planet-forming capability, and thus Class II objects (classical T Tauri stars with strong accretion from a substantial disc) provide the true 'proto-planetary' environments. These discs of Class II stars should then have suitable mass, size etc. to grow planets like those of the solar system and the extra-solar examples.

2.1 Disc masses and sizes

The Minimum Mass Solar Nebula (MMSN) must have contained at least 50 Earth masses of refractory material, as this is the summed rocky content of the giant planet cores and terrestrial planets. Scaling this up by a primordial gas-to-dust mass ratio of 100, the inferred MMSN began with at least 0.02 $M_\odot$ or 20 Jupiter masses of mostly gaseous material (Davis 2005), a large amount of which escaped since the summed masses of the planets are only ≈ 1.5 M_{Jupiter}. Since the extra-solar planetary systems so far detected are on average more massive than this, we would expect the planet-forming extra-solar discs to be more substantial than the MMSN. However, recent observations suggest that this is not the case. Andrews & Williams (2008 and earlier work referenced therein) use optically-thin millimetre dust data to find typical Class II discs to be of only 5 Jupiter masses in the Taurus and Ophiuchus star-formation regions. The situation is worsened if all stars are included (e.g. the Class III's) and if upper limits are taken into account, in which case the median disc may be only 1–2 Jupiter masses. Putting together various millimetre-wavelength surveys of nearby star clusters, less than 10 percent of stars appear to host an MMSN, which suggests that the Sun was *not* typical in its early stages but instead had one of the 'top-end' discs. However, there may still be a problem, as *more* than 10 percent of main-sequence stars in long-term Doppler surveys host giant planets. Figure 2 summarises current data on disc and planet masses – while there are various biases in detectability and targets observed, the plot illustrates that the apparent efficiency of converting disc mass into planet mass would be high, much more than the few percent in the solar system.

Most core-accretion models for forming Jupiter-analogues adopt a mass in planetesimals of a few times that of the MMSN – in order for the planet core to grow massive enough to attract the gas atmosphere, within a fast enough time that the gas-disc has not already dispersed – so this is more than an order of magnitude above the typical observed exo-disc dust-mass reservoirs. Furthermore, these models also assume a relatively high surface density, i.e. via a compact disc. This may have been true for the MMSN if we set the disc outer radius to be the ~ 50 AU distance of the Kuiper Belt, but extra-solar discs are often larger. Andrews & Williams find typical (resolved) radii of 200 AU, and so for a 20 M_{Jupiter} disc with surface density declining as $r^{-1.5}$ (as inferred from the planetary mass distribution over the solar system), the solids constitute ~ 1 g/cm^2 at Jupiter's orbit, or $\sim 5 - 10$ times lower than commonly used in models.

Since exo-planets do exist, one possibility to solve these problems is that the disc masses have been underestimated. Current avenues of research include measuring the faint tail of the dust spectrum into the radio regime – since small particles emit poorly at wavelengths longer than their own size, detecting radio emission implies cm-size grains are present. As well as being a promising sign of growth towards planetesimals, this can boost the dust (and inferred total) masses of discs, as a larger mass of low-emissivity particles is needed to reproduce the emission. A related possibility is that much of the dust could already have been incorporated into large bodies like planetary cores with small total emitting surface area, leaving an effective gas-to-dust mass ratio $\gg 100$ if the gas is still mostly in the disc. In this case, the total disc mass would be underestimated using dust data and an assumed gas-to-dust mass ratio of 100. The hot-H_2 emission from disc surfaces has been detected for a handful of discs, with models of the excitation processes suggesting of order an MMSN in underlying cold gas mass. However, the number of non-detections

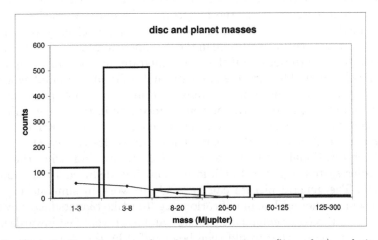

Figure 2. *Distributions by mass for planetary systems (line plot) and circumstellar discs (histogram), in log-bins of approximate width 2.5. The planetary systems (http://exoplanet.eu/) are data from Doppler surveys, summed for each host star and corrected for a mean inclination of 60°. The disc masses are from millimetre surveys in the literature of the NGC 2024, ONC, Oph, Tau, IC 348, Lup and Cha regions of star formation – each survey covered ~ 30 − 300 stars but no attempt has been made to correct for completeness or the number of stars in each cluster and upper limits are not included. For some clusters few discs were detected but a net detection was made by adding many stars together. Hence there is a large number of average low-mass discs, and a probable deficit of discs in the 8-20 Jupiter-mass bin as these are harder to detect individually.*

again implies that these may be 'top-end' discs (Ramsay Howat & Greaves 2007), leaving a picture not very different from that of the dust data.

In summary, many discs have been found but the properties appear rather different from those inferred for the disc of the young Sun. However, the optically-thin wavelength regimes correspond to cool dust typically at tens to hundreds of AU from the star, whereas planet formation takes place at a few AU. Higher-resolution data may promptly solve the puzzles of disk mass and internal mass distribution, simply by giving us a view of the small-scale material involved.

3   Effects of planet formation in discs

In a standard core-accretion model of planet formation, the first stage is that dust coagulates into planetesimals, over only $\sim 10^5$ years. Then, these bodies merge randomly to produce a planet core, in a much slower phase lasting a few 10^6 years. In this epoch, dynamical friction results in the largest bodies moving more slowly so that smaller faster bodies may collide and merge with them, and then the bulkiest can eventually attract each other by mutual gravity and merge to form the first cores, of around Mars-size. If the cores grow further up to ~ 5 Earth masses, they gain sufficient gravity to accrete gas locally from the feeding zone in the disc and can thus evolve into gas giants – with this growth boosted if the planet migrates and so can sweep up a more extended region. This process truncates naturally at the point when the gas disc disperses, so some stars may only host analogues to Neptune and Uranus with thin atmospheres, if the planets formed late in the gas era. Finally, any remnant rocky material will continue to collide and terrestrial planets may thus appear after a few 10^7 years. These timescales can be checked against both observational data such as gas-disc lifetimes, and using the history of the solar system e.g. the Earth-formation time deduced from radio-isotope dating. The general picture can be made to work in simulations, although with some rather ad-hoc modifications e.g. to make planets faster by reducing grain opacities (thus enhancing cooling), and to prevent fast inwards migration of planetary cores that would then be engulfed by the star.

An important point is that planet growth can be quite stochastic, i.e. the outcome depends on a small number of chance events. The exact 'architecture' of the solar system could thus be quite rare, and for example the number of close analogues to Jupiter is (as yet) very small in the range of extra-solar planetary systems. The Earth may also be unusual in having a large moon, which is the result of a one-off collision of the proto-Earth and a Mars-sized impactor, that resulted in a debris ring that reformed into a massive satellite. Since the Moon raises large tides, this could have astrobiological implications for the gradual emergence of life from shallow coastal waters onto dry land. Another peculiarity is that the deuterium content of water on Earth and Mars differs, indicating water that may have come from parts of the early solar system with different temperatures. For extra-solar Earths, simulations indicate they could be very much wetter or dryer than our own depending on the exact planetesimals that are involved in the planet's formation (Raymond et al. 2007). Being able to predict how many nearby stars should host life-supporting planetary systems is a very important goal involving branches of astronomy, geology and biology.

3.1   Expected outcomes

Some global outcomes are expected, even when the future planets of a young star can not be predicted exactly. Firstly, the core-accretion model of gas giant formation specifies that dust and gas should clear on different timescales, as the dust should disappear from observability first by being incorporated into planetesimals. Secondly, planet formation should clear out material locally, creating an empty ring along the planet's orbit, or a central hole in the disc if the planet is near the star. These phenomena reduce the level of emission, generally at a part of the SED related to the temperature at that orbit, since the scales of a few AU are spatially unresolved. Similarly, parts of the gas spectrum will be reduced, e.g. the line wings (Figure 1) representing fast-orbiting gas near the star. The gas and dust interact as they flow in the disc so some of these signatures are ambiguous - for example gas and small dust particles may sweep across gaps in the disc (Rice et al. 2006), flowing towards the star and then being accreted. The inner hole signature is the clearest in data so far (e.g. for GM Aur, Rice et al. 2003) – a strong decrease in near-infrared emission can best be explained by planetary clearing in the region of a few AU. Cleared rings are less likely to be identifiable from the SED, as a wide disc has grains at a broad range of temperatures that tend to fill in the spectral distribution (i.e. as a sum of blackbodies). Further, it is possible that rings could appear *without* planets, as radiation pressure forces grains out while viscosity drags other grains towards the star (Takeuchi & Lin 2003), potentially creating a natural cleared region.

The present difficulty is in imaging the effects of planet formation, since Jupiter's 10 AU wide orbit in the nearest star-formation region (the TW Hydra group at $\sim$ 50 pc) subtends only 0.2 arcsec. For atypically large discs that may have planets forming (or migrating) far out from the star, some asymmetries can just be resolved. In particular, the AB Aur disc shows both spiral-arm features in scattered light and distortions in the maps of orbiting trace molecules (Fukagawa et al. 2004). These can convincingly be explained by a perturbing planetary influence exerted at a few hundred AU – most likely by a planet that has migrated significantly outwards after having formed in the higher surface density regions of the inner disc.

At present, a large library of SEDs is available in the literature – most recently enhanced by infrared data from the ongoing Spitzer satellite mission – and so a better picture is emerging of the typical effects of planet formation on discs. For example, the number of examples of inner clearing should be much

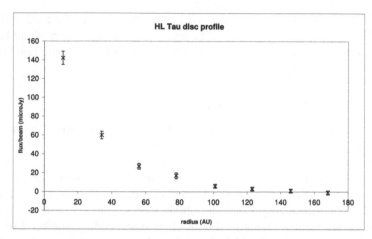

Figure 3. *Radial profile of the disc of HL Tau at 1.3-cm wavelength, observed with the VLA (Greaves et al., in preparation). Most of the emission is from large dust grains; the majority of the dust mass lies within the equivalent of Neptune's orbit (30 AU) suggesting this material can go on to form a planetary system on a scale similar to our own.*

better known when modelling of these systems is completed. The number of resolved disc images is still small, but searching for the effects of planetary perturbations is being made easier by the advent of 10m-class telescopes with cameras working in the mid-infrared. For example, the archetypal beta Pictoris disc is now seen to be warped by a potential planetary system on scales resolved down to tens of AU. Future science is thus moving towards studying the smallest scales – the $\sim$ 1-AU formation zone of terrestrial planets.

3.2   Imaging planet-formation

The most direct clues to how planets form would be obtained by imaging the process in action. Some very low-mass companions to stars have been imaged directly, including the few-Jupiter mass companion to the brown dwarf 2MASS 1207-3932 (see Mohanty et al. 2007 and references therein). However, such candidates are often argued to have formed by a process analogous to that involved in making binary star systems, since the mass ratio of the two components is small, very much less than the $\approx$ 1000 : 1 of the Sun and Jupiter. Also, the discs from which these companions formed have mostly dispersed, leaving the processes involved unknown. In the future, mid-infrared observations by JWST and millimetre imaging with ALMA will be tools to reach down to the scales of interest for observing planet formation in action. In particular, the latter millimetre interferometer, if built to the longest planned baselines and working at the highest (THz) planned frequencies, would be able to image a Jupiter analogue in formation at the distance of the TW Hydra group. Simulations (Wolf & D'Angelo 2005) produce an image with a cleared ring plus the actual planet, seen by the thermal heat-glow released as it accretes and cools. Until these facilities are built, only the radio regime offers the very high resolution needed (Figure 3). For example, the UK MERLIN array has 40 mas resolution at 5-cm wavelength, which corresponds to the orbit of the Earth at 50 pc. With the ongoing sensitivity upgrade (eMERLIN), detecting a reservoir of rocky particles in the process of forming an exo-Earth around a star in a nearby young cluster is a real prospect.

4 Summary

There is some way to go before theory and observations of planet-formation meet. In particular, mapping the large sizes and low masses of the exo-discs detected onto the mostly compact systems of heavy gas giant planets is challenging, and may require identifying only some discs as suitable – this would also place the early solar nebula in the top end of discs, somewhat against the Copernican principle. Some signatures of planet formation have been predicted that may now be searched for, and perturbations and inner holes are very promising as not being naturally produced by other mechanisms. The direct imaging of planets in formation is a real prospect over the next decade.

I have here reviewed the formation of planets from discs at early times, within the first 10 Myr of stellar life, or one-thousandth of the main-sequence lifetime for a star like the Sun. Beyond this, the formation of Earth-like planets is expected to take a few tens of Myr, and the conditions in circumstellar discs set the stage for this and some aspects of later planet habitability – these topics are reviewed in the companion chapter.

Acknowledgments

I thank the organizers and attendees of the school for making it a memorable experience (including scientifically!). Many thanks to my colleague Kenny Wood who provided a lot of the knowledge on the properties of discs.

References

Andrews, S. M., Williams, J. P. (2008), A submillimeter view of protoplanetary dust disks, *Astrophys. Sp. Sci.* **313**, 119.

Davis, S. S. (2005), The surface density distribution in the solar nebula, *Astrophys. J.* **627**, L153.

Ford, E. (2010), Dynamical evolution of planetary systems, in *Extra Solar Planets: the Detection, Formation, Evolution and Dynamics of Planetary Systems*, Editors B.A Steves, M. Hendry and A.C. Cameron, Taylor and Francis Publishing, this volume, Chapter 7.

Fukagawa, M. et al. (2004), Spiral structure in the circumstellar disk around AB Aurigae, *Astrophys. J.* **605**, L53.

Greaves, J. S. (2004), Dense gas discs around T Tauri stars, *Mon. Not. Roy. Astron. Soc.* **351**, L99.

Greaves, J. S. (2005), Disks around stars and the growth of planetary systems, *Science* **307**, 68.

Mohanty, S. et al. (2007), The planetary mass companion 2MASS 1207-3932B: Temperature, mass, and evidence for an edge-on disk, *Astrophys. J.* **657**, 1064.

Ramsay Howat, S. K, Greaves, J. S. (2007), Molecular hydrogen emission from discs in the η Chamaeleontis cluster, *Mon Not Roy Astron. Soc* **379**, 1658.

Raymond, S. N., Quinn, T., Lunine, J. I. (2007), High-resolution simulations of the final assembly of Earth-like planets. 2. Water delivery and planetary habitability, *Astrobiology* **7**, 66.

Rice, W. K. M. et al. (2006), Dust filtration at gap edges: implications for the spectral energy distributions of discs with embedded planets, *Mon. Not. Roy. Astron. Soc.* **373**, 1619.

Rice, W. K. M. et al. (2003), Constraints on a planetary origin for the gap in the protoplanetary disc of GM Aurigae, *Mon. Not. Roy. Astron. Soc.* **342**, 79.

Smith, B. A., Terrile, R. J. (1984), A circumstellar disk around β Pictoris, *Science* **226**, 1421.

Takeuchi, T., Lin, D. N. C. (2003), Surface outflow in optically thick dust disks by radiation pressure, *Astrophys. J.* **593**, 524.

Wolf, S., D'Angelo G. (2005), On the observability of giant protoplanets in circumstellar disks, *Astrophys. J.* **619**, 1114.

Wood, K. et al. (2002), Infrared signatures of protoplanetary disk evolution, *Astrophys. J.* **567**, 1183.

Debris discs and planetary environments

J S Greaves

SUPA, University of St Andrews, School of Physics & Astronomy, North Haugh, St Andrews, KY16 9SS, United Kingdom
STFC Advanced Fellow

1   Introduction

The preceding chapter reviewed the observational properties of the discs out of which planets are expected to form. The formation timescales are up to a few million years for gas giants, and tens of millions of years for terrestrial planets (based on the history of solar-system bodies). These durations occupy less than 1 % of the main-sequence lifetime of a star like the Sun (4.5 Gyr old now, age around 10 Gyr when it evolves into a red giant). Thus it might be expected that circumstellar discs are a phenemonon only of very young stars – the orbiting dust and gas being absorbed into planets or dispersed. In fact, very low-mass discs of dust particles are now known to exist around main-sequence stars, and as late as many Gyr (possibly also surviving into the giant phases).

'Debris' discs were discovered by the IRAS mission, from routine calibration observation of the magnitude-zero star Vega. The far-infrared satellite data showed Vega to be much brighter than expected, and it was quickly realised that such a far-infrared excess could be explained by orbiting dust grains, that absorb stellar light and re-emit the energy at longer wavelengths (Aumann et al. 1984). Further, since the star is not optically obscured, the dust can not be in a shell-like distribution as in giant star envelopes, but must be flatter and so most likely in a disc. For the particles to survive over very long times against drag forces and destruction in mutual collisions, it turns out that they must be continually re-generated. If the grains are in thermal equilibrium with the star then far-infrared emission implies locations at tens of AU. Hence, the interpretation of the emission is that the particles are frag-

ments produced by collisional break-up of comets – and subsequent imaging
has confirmed that the interpretation as belts of dust is correct.

After two comets collide, the fragments will undergo further collisions in a
cascade process, ending in particles small enough to be swept out of the system
by the stellar radiation pressure. This size is generally a few microns, so the
dust grains remaining in the system are large, when compared to sub-micron
sized interstellar grains and the grains commonly studied in proto-planetary
discs. Since debris is generated in a top-down process originating in larger
bodies (e.g. km-sized comets), it is a signpost that the system proceded a
good way along the road to planets at an earlier epoch.

There is thus a clear distinction between proto-planetary discs and debris discs, which is however not always appreciated in the literature. Protoplanetary discs are those seen at early times, incorporating both dust and
gas and in the growth stage of forming larger bodies. When the gas has disappeared, the orbits of solid particles are not slowed by viscous drag forces,
and so collisions start to be destructive rather than constructive. Then dust
is generated by the break-up of larger bodies, and hence although particles'
chemical compositions etc. may be similar to 'primordial' dust, the size distribution and total population is not. Observationally, the distinction is rather
clear-cut because debris disc dust masses are lower by a few orders of magnitude than the corresponding masses in proto-planetary discs (e.g. Wyatt et
al. 2003). However, 'transition' discs presumably exist, and may be identified
with the epoch when the gas mass has become very low – the β Pictoris is a
likely archetype at an age around 10 Myr. At this point both constructive and
destructive processes are probably occurring – and indeed terrestrial planet
formation should be ongoing as late as tens of Myr, from the merger of large
planetesimals, simultaneously with debris production. The term transition in
this case refers to the switch from growth to break-up phase of planetesimals
(which can be confusing since some authors use 'transition disc' to refer to
the clearing of inner holes in discs by planet formation).

The study of debris discs can yield several pieces of information about
extra-solar planetary systems. These include knowing that planet formation
commenced; the outer bounds of the system's size (where planetesimals did
not grow any more towards planets); the total populations of comets and asteroids (via the amount of collisional dust); the location of such bodies in
relation to any known planets; whether perturbations of the dust disc signal the presence of distant planets; and some idea of whether comet-planet
interactions produce a bombardment hazard for any emerging life on inner
terrestrial planets. Current research addresses these questions, starting from
such basic observational points as which stars possess debris and why; whether
the phenomenon is a steady-state or episodic; and what levels of debris exist.
For example, the Sun's Kuiper Belt of comets should appear as a very faint
debris disc if viewed externally. However, the dust mass estimated from far-
infrared emission (actually IRAS and COBE upper limits) and from impacts
of particles on spacecraft such as Pioneer indicate a very low mass of dust,

of order 10^{-5} Earth masses. This is about an order of magnitude below the emission signature that can currently be detected around other stars, even those only a few parsecs away. Thus it is difficult to place the Sun's planetary system architecture in context. Current and near-future far-infrared and sub-millimetre imaging can start to work on this problem, in particular by using very sensitive continuum cameras to reach low dust masses and resolve debris discs spatially. This gives a direct picture of where comets reside and how many exist, and if they are perturbed by Neptune-analogue planets, at least for very nearby stars. Although only about a dozen such images exist as yet, a great diversity of system architectures is seen.

2 Disc observations

Most discs have been detected first in the far-infrared, with some examples of cold discs found in the submillimetre. Warmer dust is occasionally seen in the mid-infrared, in rare systems with an inner planetesimal belt somewhat analogous to the Sun's Asteroid Belt. At longer wavelengths, emission decreases in the Rayleigh-Jeans tail of the spectral energy distribution so millimetre detections are difficult. Because of this restricted wavelength regime of the thermal emission, imaging the dust requires either satellite observations or telescopes at very high and dry mountain sites. Debris also scatters optical starlight, and this technique has recently resolved a number of discs with interesting structures, most commonly using HST.

Several limitations mean that only relatively nearby main-sequence stars can be studied effectively. Beyond about 100 pc, only systems of extremely high excess produce enough dust flux to be detected and also interstellar cirrus emission can produce false signals. For context, about 2000 star systems are known within 25 pc of the Sun, and this population offers the best chance to image discs. Typical angular resolutions are only about 10 arcsec, with either 1-3m-class satellites in the far-infrared or 10-30-m-class (sub)millimetre telescopes on the ground. Since the Sun's Kuiper Belt of approximately 100 AU diameter would then only subtend one beam at 10 pc, most of the famous well-resolved systems lie within about this distance. For context, only about 50 Sun-like stars lie within 10 pc (see the database at http://nstars.nau.edu), of which only a handful are so far known to have debris.

Consequently, only major features such as large disc cavities or substantial asymmetries can be resolved. The low angular resolution also creates a problem of confusion: particularly in the submillimetre, dusty starburst galaxies at large redshifts have similar dust temperatures and can mimic debris discs. At good survey depth, such galaxies become common on the sky and the chance that one lies behind the target star increases to a high percentage. This limits robust identification of (unresolved) discs until higher resolutions become available. Finally, the flux of very small amounts of dust – e.g. of order a lunar mass – is very small, down to the milli-Jansky regime. In the infrared this becomes small relative to the intrinsic photospheric light of the star, and then imperfect absolute calibration prevents detection of an excess. This is not a problem in the submillimetre since the star as a relatively hot and compact body produces little long-wavelength light, but then the dust emission is also reduced as it is past the blackbody peak for a few tens of Kelvin grain-temperatures.

2.1   Examples

A general impression of debris discs may be obtained from the image gallery compiled by P. Kalas.[1] Studying individual systems shows immediately that there is great diversity – and few systematic patterns. The youngest known (possibly 'transition') discs are those of β Pic and AU Mic in the same young star association, around 10 Myr old, but of A and M spectral type respectively. Both discs are asymmetric suggesting perturbations by newly-formed planets – in particular β Pic has tilts of the inner disc plane that are difficult to explain without a planetary system. Both discs are large, at hundreds of AU in radius, possibly reflecting recent development out of proto-planetary discs.

Later examples include the well-studied stars Vega and Fomalhaut, both A-type objects of order 200-300 Myr old; however, the discs differ greatly. Vega is unusual – although the archetype – because it has a large far-infrared dust halo interpreted as blown-out grains. The effects of observing at different wavelengths are notable: in the (sub)millimetre clumpy structure appears, and this has been interpreted as grains trapped in orbital resonance, by a distant giant planet as it migrated outwards. (Since trapping efficiency and emitting wavelength are both related to grain size, it is necessary to be aware that disc appearance does vary with observing wavelength and is not 'what you see is what you get'.) Fomalhaut has no such halo and appears as a thin ring of dust, 'limb'-brightened at the ends as it is seen fairly edge-on and so the ends possess a greater column length of optically-thin dust. It has been noted in both the submillimetre and optical that the centre of the dust ring is offset from the stellar position, and this can be explained by forcing of the dust particles' orbits by an eccentically-orbiting planet. The dust ring also has an asymmetry in emission suggesting a very distant planet out at around 100 AU, which would be at a remarkable three times the size of the orbit of Neptune.

Somewhat later in age, debris has been imaged (Figure 1) around the K2 star ε Eridani. Because it is very close, the stellar radius is measurable with stellar interferometry, and combined with models of size changes as nuclear burning proceeds along the main sequence, this yields a robust stellar age (di Folco et al. 2004). At 850 Myr, the star is similar to the Sun at the end of the Late Heavy Bombardment episode (around 700 Myr) and the image supports a similar history. The centre is partly cleared out while a large colliding population of planets remains, about 50 % further out than the Kuiper Belt. A

---

[1]Currently hosted at http://astro. berkeley.edu/ kalas/discsite/pages/gallery.html.

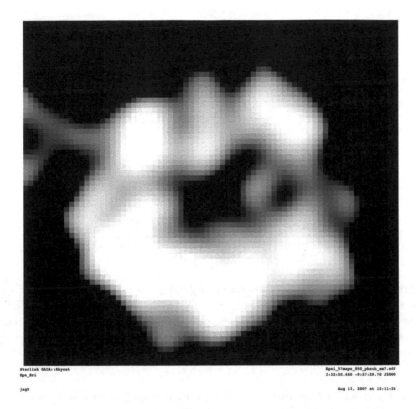

Figure 1. *Image of the ϵ Eridani debris disc obtained with the SCUBA submillimetre camera on the James Clerk Maxwell telescope. The grey scale shows flux at 850 microns wavelength and the dust ring emission peaks at 65 AU; the star is 3.2 pc from the Sun. This is the 9th closest stellar system to the Sun and also possesses a gas giant on an eccentric orbit at about 3 AU semi-major axis.*

complex system of clumps points to a perturbing planet. Further, the ring has been observed long enough that there is tentative evidence of rotation of the features (Poulton et al. 2006) – if confirmed this will tie down the period and so semi-major axis of the planet that enforces the resonant-clump pattern.

Beyond the age of the Sun, τ Ceti has been imaged in the submillimetre – this is again a nearby star with a radius measurement, and an age of close to 10 Gyr, twice the Sun's present age. The population of comets is more than an order of magnitude (in mass) greater than around the Sun, in spite of the greater age of τ Ceti, one of the earliest Galactic stars. The large planetesimal population, combined with the lack of a 'shielding' giant planet (no Doppler signal in spite of many years of monitoring), suggests that any terrestrial planet in the system could have undergone massive cometary impacts for the whole stellar lifetime. Any life here would in this case have had to evolve very

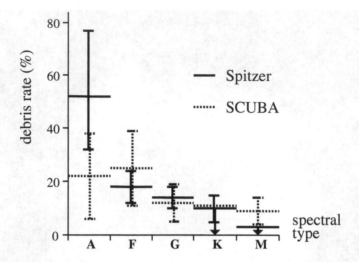

Figure 2. *Incidence rates of debris discs from surveys of main-sequence stars in the mid-to-far-infrared (Spitzer, ISO, IRAS) and submillimetre (SCUBA). Error bars are from Poisson statistics, i.e. from $\sqrt{N_{stars}}$. Note the differences between the wavelength regimes, probably explained by sensitivity to warm versus cool dust (see text).*

differently from that on Earth, perhaps underground or in the oceans rather than emerging onto the hazardous land environment.

2.2 Trends

Many more debris systems have been discovered recently from far-infrared excesses measured by the Spitzer satellite, down to levels of about 50 % above the photosphere. While the Sun's Kuiper belt excess would be only ~ 2 % or less, these data still reach down into a new population of moderate amounts of comets. A few trends are emerging. Firstly, the incidence of debris around A-stars is around 50 %, compared to 15 % for FGK stars and very few detections for M-stars (Figure 2). In part, this is explicable by evolution with age, as mid-A-type stars last only about 1 Gyr on the main sequence, and so it is not surprising that they are often seen in the equivalent of the heavy bombardment era. The low incidence for M-stars may be an observational bias, as these stars are of low luminosity and so in thermal equilibrium the dust will be cold – this interpretation is supported by a rate ~ 10 % from submillimetre surveys, more like that of Sun-like stars. Secondly, the incidence is generally higher for ages of a few hundred Myr, but does not decline greatly with time over the 1 to 10 Gyr epoch. This suggests that the comet belts survive rather than being substantially ground down to dust, but it is not yet explained why some stars have debris and others do not, or if the phenomenon could be universal but episodic. Thirdly, it is not possible yet to predict whether a star will have debris based on readily measurable stellar properties, with no relation to metallicity or rotation for example. Age is a factor, and stellar multiplicity is also emerging as a relevant property – higher rates (Trilling et al. 2007) suggest that perturbations of a companion star may stir up more planetesimal collisions.

From imaged systems (possibly biased towards the large end of the disc population), a few more trends have emerged. The dust is generally located in a ring rather than a filled disc; this central clearing can be from collisional grain destruction or ejection by a planet or both. In unresolved systems, the general lack of mid-infrared emission also implies that central cavities are common. The majority of the resolved discs show asymmetries of various kinds, including warps, offsets and clumps, and these are hard to explain without some kind of perturbing object. Finally, many discs are larger than the solar system (modulo the possible bias), with outer radii up to about 300 AU. This has implications for planet formation via the primordial disc surface density, as noted in the previous chapter.

3   Theoretical interpretation

Some general features of debris evolution have been proposed, based on the observational data. The first Gyr of stellar life is expected to be the most active, owing to late migration of planets that stirs the planetesimal orbits, and to collisions of large bodies in the last stages of planet formation. At later times, there may be a steady state with a low rate of random collisions among the planetesimals. However, the moderate incidence of debris among few-Gyr solar analogues, at levels above the dustiness of the solar system, suggest perturbations can still be ocurring. One possibility is that bodies of Pluto-like size may be still be forming slowly at tens of AU from the star, and they then cause a delayed stirring effect on the comet belt (Dominik & Decin 2003). This is surprising since we are not used to thinking of 'planet formation' still ongoing at Gyr ages! However, simulations (Kenyon & Bromley 2004 and subsequently; Thébault & Augereau 2007) imply that this can in effect cause waves or outwards-moving rings of late dust production. If this is correct, then the low-dust state of the solar system would be a function of its rather small size, with the growth stage reaching the outer bound at around 50 AU relatively early.

The general models of Wyatt et al. (2007) take an analytical approach to predicting debris as a function of time, varying such properties as stellar mass, initial disc mass and disc size, and can be tested by observed properties such as debris mass at different times. The broad range of amounts of debris for similar stars is likely to be explained by the variation among initial properties including disc mass and size, for the former of which no independent relic information exists in main-sequence stars. Given the wide range of mass and size observed among proto-planetary discs, however, it is not surprising that debris properties should have extreme variations.

4 Planetary signatures

Various clues to planets exist in the debris data. Firstly, since the dust is generated by a top-down collisional process, planet formation for this star must have reached at least the planetesimal stage. The largest size of a body that should have undergone one collision at a specified stellar age may be estimated assuming the fragments form a collisional cascade. For mature main-sequence stars this size is a substantial tens-of-kilometres – a good pointer that planet formation could have gone on to be successful. Secondly, the location of a debris belt provides an outer bound (beyond which planetesimals presumably did not form efficiently) to the scale of the planetary system. Only about ten systems are known with both a debris disc and a giant planet, and notably these planets orbit relatively far out (not hot Jupiters). Hence, the scale or mass of the disc may have dictated some aspects of the planetary configuration. Thirdly, disc structures such as warps, spiral arms, offsets and clumps all point to interaction of dust and planets. Compact clumps produced by gravitational resonance can even provide a map to pinpoint the location of a planet. A clear example is the pair of unequal 2:1 resonance clumps seen in the Vega disc, with a simple match to dust trapping models. The inferred planet is of Neptune-like mass (the mass affects the trapping efficiency and so the brightness of the clumps seen) and now orbits at approximately 65 AU – far beyond the distances that can be probed by other techniques such as Doppler wobble. Tracking the rotation of clumps has been pioneered for ϵ Eridani, and is a technique for the future that can directly confirm such planets.

Debris discs are thought to play a role in planetary migration – typically in the outwards direction rather than the predominantly inwards mode related to hot Jupiters. Angular momentum can be exchanged between a planet and planetesimals in a ring outside it, so that the planetesimals move in while the planet moves out. The cometary bodies may then cross a radius in resonance with the planet position, which results in stirring and changes to the comets' orbits. Such an event in the solar system (Tsiganis, 2010, Chapter 8) is thought to have produced the Late Heavy Bombardment when many infalling comets hit the Earth and Moon at an age of around 700 Myr. Looking for such episodic debris episodes in extra-solar systems can shed light on the later stages of planet building. For example, in clusters of stars up to $\sim$ 100 Myr old, the rate of Spitzer debris detections is a few tens of percent (Siegler et al. 2007) and debris incidence remains higher up until about 1 Gyr compared to older Sun-like stars. This suggests migration and re-arrangement of planetary systems over timescales similar to that of the Earth's construction and later resurfacing by bombardments.

5 Astrobiological implications

If the high rates of debris at a few tens of Myr reflect a colliding population of planetesimals extending up as far as bodies of Mars-like size, then terrestrial planets may be forming by merger events. The masses of dust in primordial discs are enough to form Earth-mass planets, in the minimum sense of 1 $M_\oplus$ of rocky material, for about two-thirds of stars based on millimetre surveys of young star-forming regions. These two pieces of evidence suggest that stars with some kind of Earth-analogue may be common. The major astrobiological question then is, could many of these planets be habitable?

Studies of debris shed an interesting light on this. For example, comets can deliver vital water to a terrestrial planet via impacts. However, frequent impacts of comets of size around 10 km would be a severe hazard to life - one such event 65 Myr ago may have contributed significantly to environmental changes and the extinction of the dinosaurs. This was not entirely catastrophic since mammals including ourselves then emerged, but much more frequent impacts could be a severe problem. After such environment changes, biological studies show that the number of species rises to prior levels only after $\sim$ 10 Myr. Since in the known debris systems, the dust levels reflect comet populations orders of magnitude higher than in the solar system, the impact frequency could also be very much higher. In this case, land life seems unlikely – for example, the planetary surfaces could be permanently molten – and so aquatic life-forms might be the norm.

However, much more work needs to be done on the effects of giant planets lying between the Earth zone and the comet belt. Jupiter is efficient in flicking away inbound comets that might otherwise impact the Earth, but different giant planets may have other effects. Recent work (Horner & Jones 2008) suggests a lower-mass gas giant might drop even more comets on our heads; conversely with no giant at all the comets might orbit unperturbed in a distant belt and not fall inwards. Future modelling can investigate whether debris discs in conjunction with known exoplanets create good or bad environments for life on Earth-like planets.

6 Summary and future

Since the discovery of debris discs over 20 years ago, and especially in the last decade of imaging, a number of surprises have emerged. The scale of planetary systems is often larger than our own, both in the size of the comet belts and the orbits of giant planets inferred from perturbations. The populations of comets are much higher than in the solar system, sometimes for stars that are otherwise similar to the Sun. The reasons for high levels of debris are still not well understood, but may emerge soon with the next generation of surveys. In particular, 2008 brought the expected launch of the Herschel far-infrared satellite and first observations with the SCUBA-2 submillimetre camera. Several hundred hours of time has been allocated to both unbiased debris surveys and searches towards Sun-like stars that are prime targets for exo-Earths. By careful comparison of rates and masses of debris with stellar properties such as age, type and binarity, it is hoped that the origins of debris discs will emerge. Further, we can probe down close to Kuiper Belt levels of dust and so begin to find systems like our own in both planet and comet-belt terms, that may host similar planets to the Earth.

Acknowledgments

I thank my many colleagues whose work has contributed to the results discussed here, in particular the SCUBA team for making so much of the imaging possible and Mark Wyatt for many mind-expanding discussions.

References

Aumann, H. H. et al. (1984), Discovery of a shell around α Lyrae, *Astrophys. J.* **278**, L23.

Di Folco, E. et al.(2004), VLTI near-IR interferometric observations of Vega-like stars. Radius and age of α PsA, β Leo, β Pic, ϵ Eri and τCet, *Astron. Astrophys.* **426**, 601.

Dominik, C., Decin, G. (2003), Age dependence of the Vega phenomenon: theory, *Astrophys. J.* **598**, 626.

Horner, J., Jones, B. W. (2008), Jupiter: friend or foe?, *Astron. Geophys.* **49**, 22.

Kenyon, S. J., Bromley, B. C. (2004), Collisional cascades in planetesimal discs. II. Embedded planets, *Astron. J.* **127**, 513.

Poulton, C. J., Greaves, J. S., Cameron, A. C. (2006), Detecting a rotation in the ϵ Eridani debris disc, *Mon. Not. Roy. Astron. Soc.* **372**, 53.

Siegler, N. et al. (2007), Spitzer 24-micron observations of open cluster IC 2391 and debris disc evolution of FGK stars, *Astrophys. J.* **654**, 580.

Thébault, P., Augereau, J.-C. (2007), Collisional processes and size distribution in spatially extended debris discs, *Astron. Astrophys.* **472**, 169.

Trilling, D. E. et al. (2007), Debris discs in main-sequence binary systems, *Astrophys. J.* **658**, 1289.

Tsiganis, K. (2010), Late stages of Solar System formation and implications for Extra-solar systems, in *Extra Solar Planets: the Detection, Formation, Evolution and Dynamics of Planetary Systems*, Editors B.A Steves, M. Hendry and A.C. Cameroon, Taylor and Francis publishing, this volume, Chapter 7.

Wyatt, M. C., Dent, W. R. F., Greaves, J. S. (2003), SCUBA observations of dust around Lindroos stars: evidence for a substantial submillimetre disc population, *Mon. Not. Roy. Astron. Soc.* **342**, 876.

Wyatt, M. C. et al. (2007), Steady state evolution of debris discs around A stars, *Astrophys. J.* **663**, 365.

Dynamical evolution of planetary systems

Eric Ford

University of Florida; Harvard-Smithsonian Center for Astrophysics; Hubble Fellow

1 Planet formation

The modern planet formation literature can be roughly divided into two frameworks for the formation of giant planets.

In the core accretion model, collisions of dust grains and then small rocky bodies result in the gradual accretion of a rocky core. If the core reaches a critical mass before the gas has dissipated, then it becomes a gas giant by accreting a large envelope of gas. The early stages of this model involve complex chemistry and physics that are poorly understood. Additionally, there are concerns related to how rapidly this model can produce the core needed to capture the gas from the early protoplanetary disk.

In the disk fragmentation model, an entire gas giant planet collapses directly from the gas in the protoplanetary disk. The primary feature of this model is that such instabilities could develop very rapidly, while there is still plenty of gas available. While some simulations appear to show such instabilities, one should pay attention to several details of these models, such as the choice of initial conditions, the boundary conditions, the equation state for the gas, the cooling time, and the duration of the simulations. Recent work has cast considerable doubt as to whether disk fragmentation can indeed form Jupiter-mass planets from the disks commonly observed around young stars (Rafikov 2005).

In weighing the core accretion versus disk fragmentation "debate", one should keep in mind that no one seriously suggests that disk fragmentation could form the terrestrial planets. While the details of core accretion are unclear and worthy of the significant ongoing research, it appears to be the only viable model for the formation of terrestrial planets. Given the lack of any credible alternative, the existence of the terrestrial planets and structure of meteorites provide clear evidence that protoplanetary disks can create rocky cores via accretion. Therefore, there is a certain economy of hypotheses in assuming that core accretion also accounts for the formation of the cores of giant planets. In the remainder of this chapter, we will ground our discussion in the framework of the core accretion model.

1.1 Constraints from observations of protoplanetary disks

Key phases of planet formation are notoriously difficult to observe. Planet formation occurs around young stars which are rare compared to main sequence stars and therefore more distant. The same gas, dust, and planetesimals responsible for planet formation often obscure the young stars that provide the photons necessary for study from Earth. Further, planet formation occurs on such small length scales that direct observations of planet formation are well beyond the capabilities of current observatories. Most observations of planet-forming disks could be placed into one of three categories.

- Observations of the spectral energy distribution star and any surrounding gas, disk, and planets. After modelling and subtracting the stellar emission, the remaining emission at each wavelength is generally interpreted as being dominated by particles at a given equilibrium temperature and hence distance from the star. A combination of two local maxima in the SED is considered to be the characteristic signature of a debris disk where the small bodies responsible for the infrared emission are being replenished by a collisional cascade. In recent years, near and mid-infrared observations by both Spitzer and ground-based observatories have greatly advanced this research (Meyer et al. 2006; Andrews & Williams 2007).

- Imaging at optical, near-infrared, and sub-millimeter wavelengths can provide valuable information about disk morphology and structure. Disk edges and asymmetries (e.g., warps, clumps, offsets) appear to be common and are often interpreted as originating due to gravitational perturbations of unseen planets or protoplanets. Images in optical light are dominated by scattering of star light toward the Earth by small grains. On the other hand, images at mid-infrared or sub-mm wavelengths are due to the thermal emission of small bodies, so different wavelengths probe material with different sizes, temperatures, locations in the disk. Therefore, considerable care is necessary in interpreting detections and non-detections at various wavelengths.

- Interferometric observations are capable of resolving the inner regions of nearby disks. Such observations have been particularly useful for measuring the distance to the inner clearing of the disk (Eisner et al. 2007).

Again, interpretation is delicate and the focus of significant current research.

Clearly, there is a rich and interesting set of disk observations that is worthy of a much more extended discussion (see Chapter 5 by Greaves in this volume). Here we extract only a few key points from the existing observations.

- Gas-rich, primoridal disks containing primordial ISM dust grains last for an order of one million years.

- Debris discs contain little to no gas, but significant amounts of dust that must be replenished by collisional cascade. Such disks appear from one to ten-million years.

- The transition from primordial disk to debris disk occurs quite rapidly.

- The typical mass of metals in potentially planet forming disks is comparable to that believed to have formed the planets in our solar system.

These observations lead to several challenges which must be addressed by planet formation theory. First, giant planets dominated by gas (outer planets in our solar system, transiting "hot Jupiters" around nearby stars, and likely most of the giant planets found via radial velocity planet searches) must have formed in a disk that still contained a significant amount of gas. Given the short timescales for dissipation of the gas around other stars, this places a significant constraint on the timescale for the accretion of a massive rocky core for Jupiter and Saturn, and very strong constraints on the formation of Uranus and Neptune. Indeed, this timescale problem has often been cited as a potential weakness of the core accretion model. Therefore, considerable research has been focused on understanding how various physical processes might result in accelerated accretion timescales (compared to more simplistic models).

1.2   Classical planet formation via sequential accretion

First, a protostellar nebula containing gas and small interstellar dust grains collapses to form a protostar and gas-rich disk. Dissipation in the disk leads to accretion onto the star and heats the inner regions of the disk. In the inner solar system primordial grains can be destroyed and new dust grains condense as the disk cools. Both the dust and ice grains are tightly coupled to the gas disk, but settle towards the midplane of the disk. Once concentrated (commonly thought to occur in the midplane, but perhaps in other locations such as vortices or random overdensities created by turbulence; e.g. Barranco & Marcus 2005), these dust grains begin to coalesce into large bodies. Direct collisions might be able to produce pebble-sized bodies, if the relative velocities are sufficiently small to prevent the shattering into smaller bodies. Alternatively, a small scale near the midplane might lead to conglomerations the size of boulders, bypassing the pebble-sized regime, where a rapid inward migration could prevent further mass growth (Goldreich & Ward 1973).

In either case, once the bodies reach boulder sizes, they are only weakly coupled to the gas, and they enter a regime where the physics (gravity, accretion of rubble piles, minor gas drag) is better understood. At this point, the very large number of bodies makes direct n-body simulations impractical, but also makes it practical to apply the principles of statistical mechanics (e.g., Folker-Planck or collisional Boltzman equation). Collisions leading to both mergers and fragmentation result in a gradual growth of planetesimals including a broad distribution of sizes and a tail that extends to tens to hundreds of kilometers in radius.

As the gravitational stirring by the bodies in the large mass tail of the distribution becomes significant, it becomes important for simulations to explicitly account for the small number of more massive bodies. Since most of the mass is still in smaller and much more numerous bodies, the tools of choice are hybrid codes that include both a statistical treatment of small bodies and an n-body integrator to account for the gravitational perturbations of the more massive planetesimals and/or protoplanets (Bromley & Kenyon 2006). At this point, the growth rate of the most massive protoplanets exceeds that of the typical planetesimal, resulting in one (or possibly a few) protoplanets in each of many annuli, where each protoplanet is well-separated in space from the other protoplanet and well-separated in mass from the bulk of planetesimal mass function within its annulus (Kokubo & Ida 2002).

As the protoplanets reach roughly a lunar mass, the random velocities of the smaller planetesimals are excited by the gravitational interactions with the more massive protoplanets in the same and neighboring annuli, thus reducing the efficiency of gravitational focusing and the rate of accretion. Since the rate of stirring increases as the mass of the protoplanets increases, the growth rate of the most massive protoplanets slows, so that less massive protoplanets have an opportunity to "catch-up" with more massive protoplanets in other annuli. At the end of this phase of classical oligarchic growth, each protoplanet reaches the isolation mass that is set by the disk surface density and the spacing of the oligarchs (Goldreich et al. 2004; Chambers 2007). If the protoplanetary core reaches a critical mass of $\sim 10 M_\oplus$ soon enough (while the disk is still gas rich), then the core can begin accreting a potentially much larger mass of gas to become a gas giant.

1.3   Chaotic phase of planet formation

The recent discovery of diverse extra-solar planetary systems strongly suggests that planetary systems continue to evolve on even longer time scales. While collisions and gas drag are not efficient for directly damping the random velocities of the more massive protoplanets during oligarchic growth, these same forces can dampen the inclinations and eccentricities of the planetesimals. As a result, the planetesimals maintain very small random velocities, so that they provide strong dynamical friction that can limit the eccentricities and inclinations of the protoplanets as they grow. Alternatively, if significant gas remains, then it may contribute to preventing close encounters among the protoplanets during this stage. Regardless of whether the dissipation is dominated by gas or dynamical friction, the rate of eccentricity damping is limited by the total mass in the disk, as opposed to the mass in oligarchs.

The regular spacing between oligarchs is set during an era when there is still significant mass accretion and hence damping of the oligarchs' random velocities. As this material is accreted and the sources of dissipation are removed, the mutual stirring of the oligarchs will lead to increasing eccentricities and eventually close encounters between oligarchs (Goldreich et al. 2004; Ford & Chiang 2007). Additionally, migration due to the gravitational scattering of planetesimals may cause late stage orbital migration of the oligarchs. Decreased separations between the oligarchs may directly lead to stronger planet-planet interactions, or the migration may lead to the divergent crossing of mean-motion resonances that can trigger violent instabilities among the oligarchs and throughout the disk (Tsiganis et al. 2005).

2 New wrinkles to planet formation theory

In reality, planet formation is even more complicated. Even if theorists were once tempted to envision a relatively simple formation scenario for our own solar system, the discovery of giant planets around nearby stars has demonstrated that additional processes can often dramatically influence the final state of planetary systems. The existence of short-period giant planets provides evidence supporting large scale orbital migration. The existence of giant planets with large orbital eccentricities demonstrates that there must be at least one common eccentricity excitation mechanism that can induce large eccentricities of Jupiter-mass planets. While theorists have proposed viable mechanisms to explain each of these observations, it is not yet clear which specific mechanisms are most important for sculpting planetary systems in general. There is also significant research devoted to understanding the extent to which these processes operated in the solar system. Some of the proposed mechanisms (e.g., migration in a gas disk) must be incorporated into simulations of the early and mid-stages of the planet formation processes. Other proposed mechanisms (e.g., migration in a planetesimal disk and late-stage dynamical instabilities) for orbital migration and eccentricity excitation would operate on timescales extending well-beyond the era of planet formation and can be visualized as adding on another stage of dynamical evolution after the end of planet formation.

2.1 Orbital Migration

Theoretical arguments predating the discovery of extra-solar planets suggest that planet-disk interactions in a quiescent disk would result in rapid migration. This poses two significant challenges to the classical theory of planet formation. First, small rocky planetesimals should rapidly spiral into the star (Goldreich & Tremaine 1980). This suggests that small planetesimals grow very rapidly (e.g., perhaps via a gravitational instability in the planetesimal disk the Goldreich-Ward mechanism). The second challenge is that giant planets should also spiral into the star via gas inspiral on timescales shorter than the lifetime of the gas disk (Ward 1997). Prior to the discovery of extra-solar planets, such theoretical arguments for rapid migration were largely dismissed, since they "obviously" were not relevant for our solar system.

In light of the discovery of extra-solar planets with unexpected orbital properties, it is worth revisiting the classical model for the formation of the solar system. This process is still ongoing and scientists are only beginning to work towards a new consensus model for our solar system. For example, let us consider the possibility of significant migration of giant planets in our solar system. Did the solar system once harbor additional planets that migrated all the way into the Sun? If the Sun had accreted a giant planet's worth of material enriched in metals after it had developed a thin convective envelope, then the solar core would have a lower metal content than observed in the solar photosphere. Helioseismology and neutrino measurements can place limits on such models. However, if the Sun had accreted planets shortly after the formation of the sun, while it was still fully convective, then the metals would be mixed throughout the Sun. Therefore, it may be impossible to recognize any evidence of such accretion with observations of the sun (Ford et al. 1999).

Next, let us consider the possible migration of the current giant planets. On one hand, the width of gaps in the asteroid belt suggests that Jupiter has not undergone significant migration since the formation of the asteroid belt. On the other hand, the structure of the Kuiper belt (i.e., abundance and orbital elements of Kuiper belt objects, particularly those in mean-motion resonances) strongly suggests a significant outward migration of Neptune. Dynamical studies show that scattering of a planetesimal disk by the outer planets could simultaneously produce both the needed migration of Neptune and lack of migration of Jupiter. Further, the large time-scale for *in situ* formation of Uranus and Neptune suggests that they are likely to have formed closer to the Sun and migrated outwards. In principal, such a migration could be either a slow and smooth (e.g., due to repeated gravitational scattering of many small planetesimals; Hahn & Malhotra 2005) or it might begin with the more rapid and violent gravitational scattering of planet-mass bodies (Thommes et al. 1999).

Recently, it has been suggested that perhaps an outward migration of Uranus and Neptune is not needed to solve the problem of their apparently long formation timescale. If Uranus and Neptune were able to accrete much of their mass very efficiently from very small bodies, then perhaps they could form *in situ* before the gas in the protoplanetary disk was depleted. However, based on the sequential accretion theory and the masses of Uranus and Neptune, the solar system would be expected to have formed a few additional Neptune-mass planets that are not currently observed. To match present observations, these additional Neptune-mass planets would have to have been ejected from the solar system. Such strong excitation of these now-lost planets would inevitably also excite the eccentricities of Uranus and Neptune (Goldreich et al. 2004). In this model, the observed eccentricities of Uranus and Neptune would imply that significant dissipation has occurred since the time when the additional oligarchs were ejected. Since only a small amount of gas could remain at that time, the source of dissipation is most likely the scattering of planetesimals. The scattering of the additional oligarchs (as well as a significant mass of planetesimals) would result in an outward migration of Uranus and Neptune. Therefore, even in models with mass growth dominated by highly efficient accretion of small bodies, Uranus and Neptune would need to have formed significantly closer to the Sun in order to match the current masses and orbits of the giant planets in the solar system (Ford & Chiang 2007; Levison & Morbidelli 2007).

The discovery of short-period giant planets clearly led to a revival of planet-disk migration theories. As the field matures, the extra-solar planetary systems are beginning to tell a more complex story. In many extra-solar planetary systems, giant planets on highly eccentric orbits suggest that these planets have not been subject to strong dissipation, at least since the time they acquired their eccentricities. Other extra-solar planetary systems with multiple planets in mean-motion orbital resonances that suggest a previous phase of orbital migration in the presence of significant dissipation. If stars typically form several planets at roughly comparable distances, then smooth convergent migration (e.g., in a smooth quiet disk) would be expected to frequently produce resonant planetary systems (Nelson & Papaloizou 2002). On the other hand if orbital migration is divergent and/or uneven (e.g., due to turbulence in the disk), then systems can pass through or jump over mean motion resonances and resonant trapping might occur more rarely. Theorists are actively pursuing both analytic and numerical research to better understand the details of orbital migration, the rate of orbital migration, the frequency of resonant trapping, and how migration can be halted (e.g., Trilling et al. 2002; Dobbs-Dixon et al. 2007). As migration models mature to the point where they can make definitive predictions, it will become possible to make more powerful comparisons between their predictions and the observed distributions of extra-solar planet masses and orbits (e.g., Ida & Lin 2000; Armitage et al. 2002).

2.2   Eccentricity excitation

The surprising discovery of giant planets on eccentric orbits has also provoked significant theoretical research. Again, theorists have proposed several mechanisms that could excite such eccentricities. We can crudely divide these mechanisms into two categories: mechanisms that are inevitable in at least some systems and mechanisms which might or might not operate in actual planetary systems. In reality, there is a continuum with some mechanisms (e.g., strong dynamical interactions of multiple planet systems) falling between these two extremes, due to uncertainties in the "initial conditions" that are provided by the earlier stages of planet formation.

2.3 Passing stars and wide binary companions

Inevitable eccentricity excitation mechanisms are based on well-understood physics and initial conditions that are known to occur after the most uncertain stages of planet formation have passed. For example, planetary systems that form around stars in dense stellar clusters can be perturbed by gravitational interactions with passing stars (Adams & Laughlin 2003; Zakamska & Tremaine 2004). Similarly, planetary systems that form around one member of a wide-binary star can be influenced by the secular perturbations of the distant star (Holman et al. 1997; Ford et al. 2000). While these mechanisms inevitably affect some planetary systems, they can not explain the observed high frequency of eccentric giant planets. Passing stars are only significant for stars born in dense clusters and only until the cluster dissolves. Therefore, most planetary systems would not be significantly excited by this mechanism (Adams et al. 2006). Even if we assume that those planets with host stars having no known binary companions actually do have faint and/or unresolved binary companions, the distribution of eccentricities does not match that predicted by secular perturbation from wide stellar binaries. In particular, systems with low (high) relative inclinations tend to result in low (high) eccentricities. As a result, the perturbation by a wide binary companion model is unable to reproduce the high abundance of giant planets with intermediate eccentricities (Takeda & Rasio 2006).

2.4 Planet-planet interactions

2.4.1 Strong planet-planet scattering

If the typical protoplanetary disk forms multiple giant planets, then the gravitational interactions of the planets can lead to large eccentricities. In the simplest scenario, two or more giant planets begin to form in two nearby regions of the protoplanetary disk. While the planet masses are still small, the planet-planet interactions are negligible. While the planets are accreting from a disk, the disk will also provide a mechanism for damping random velocities, limiting eccentricity growth, and preventing close encounters. As the planets accrete most of their mass and the disk mass becomes negligible, eventually the mutual planetary perturbations will lead to eccentricity growth. If the planets are sufficiently closely spaced, then the orbits will eventually cross, leading to a violent instability with close encounters and eventually planets colliding, being ejected from the planetary system, and/or being scattered into the central star. For a system with only three massive bodies (one star and two planets), there is a sharp transition (in terms of the initial spacing) between systems that are provably Hill stable (no close encounters) and those that will rapidly result in a violent instability. Even so, a modest fraction of such systems would persist for well over $\sim 10^6$ orbital times before a planet is ejected from the system (Ford et al. 2001). For planetary systems with three or more planets, the transition from (empirically) stable to unstable is more gradual (again, in terms of the initial spacing). Even if each pair of planets in isolation would be provably Hill stable, the combined planetary system will often result in violent instabilities, though on much longer timescales (Chambers et al. 1996; Chatterjee et al. 2007).

Since the physical processes that determine the masses and initial separations of protoplanets are distinct from those that eventually result in dynamical instabilities, the early stages of planet formation should not be expected to result in planet masses and orbits that will be stable for billions of years. As long as planet formation produces systems with multiple planets, we should expect that the typical outcome will be systems that are *unstable*. Initially, one pair of planets with a relatively short timescale to instability will begin to undergo close encounters. The close encounters will strongly perturb the orbits of both planets and may also propagate throughout the planetary system. Following a period of strongly chaotic evolution, one or more planets will collide, be ejected from the planetary systems, and/or be scattered into the central star. The resulting planetary system will have a reduced number of planets with larger spacing between their orbits. The systems will often undergo a stage of relatively mild chaotic evolution, where close approaches do not occur for some time. However, there is still a strong chance that the system will undergo yet another instability on a longer timescale. As long as three or more massive planets remain, the typical timescale for orbital instabilities will increase so that it approaches the age of the system. Dynamical simulations with a wide variety of initial conditions have shown that planetary systems with many planets typically relax to a "final" state of $\sim$1-3 giant planets in eccentric orbits (Lin & Ida 1997; Papaloizou & Terquem 2001; Adams & Laughlin 2003; Juric & Tremaine 2007).

2.4.2 Mean-motion resonant interactions

In principle, multiple planet systems can result in strong eccentricity excitation, even in the absence of close encounters. In particular, resonant interactions can efficiently pump up the eccentricities of well-separated planets (Lee & Peale 2002). Some theorists have suggested that a planet forming in one region of the disk can induce the formation of a second planet near a 2:1 mean motion resonance (Armitage & Hansen 1999). In the absence of such a mechanism, planets would only rarely form in such resonances. However, if there is significant orbital migration, then planets that formed outside of resonance could often pass through or become trapped in mean motion resonances like the 2:1 and 3:1 (Nelson & Papaloizou 2002). The migration might be driven by either a gas or planetesimal disk. Given the challenges of modelling even a single planet migrating through a gas disk, further research on the interactions of giant planets migrating through a gas disk is in order. The study of multiple planets migrating through a planetesimal disk is also challenging, but tractable with modern computational resources (Levison & Morbidelli 2007). While significant progress has been made in the context of the outer solar system, there is a large parameter space that remains to be explored (Ford & Chiang 2007).

2.4.3 Secular interactions

Secular interactions among multiple planet systems can lead to significant orbital evolution. In the secular approximation, the semi-major axes of each planet remain constant, while the eccentricities and inclinations are free to evolve. For systems with nearly coplanar orbits, eccentricities can be transfered between planets, but a wholesale increase in eccentricities is prevented by the conservation of angular momentum. For many systems, additional perturbations (e.g., precession due to general relativity, stellar oblateness, and/or a remnant disk) will limit the exchange of angular momentum between planets, but in some cases a resonance (between the precession rates) may allow the perturbations to grow for an extended period of time (Ford et al. 2000; Adams & Laughlin 2006).

2.5   Additional proposed mechanisms

While it seems inevitable that the above mechanisms will play some role in shaping the orbits of planetary systems, there are additional possible mechanisms that are somewhat more speculative. For example, Namouni (2005) has recently suggested that strong and asymmetric stellar jets might be capable of exciting eccentricities. As another example, planet-disk interactions offer another possible mechanism for exciting orbital eccentricities (Goldreich & Sari 2003). However, there is not yet a consensus as to whether these mechanisms will work in practice. Currently, it seems that planet-disk interactions are most likely to excite eccentricities for very massive planets $\geq 10 M_{\rm Jup}$ (Artymowicz 1992). While current observations do not dictate a need for these more speculative mechanisms to be effective, existing observations do not rule them out, either. At the moment, it seems that more theoretical work is needed to determine when (if ever) planet-disk interactions will lead to eccentricity excitation, rather than damping.

3   Multiple planet systems

Theoretical research predicts that mature planetary systems typically contain 2 or 3 giant planets (Adams & Laughlin 2003; Juric & Tremaine 2007), and radial velocity surveys suggest that at least 30-50% of exoplanet host stars already show some evidence of additional companions (Wright et al. 2007). Most current planet detections are based on observing the perturbations of planets on the radial velocity of their host stars (Butler et al. 2006). A planet is said to be detected when observations are inconsistent with a no-planet model, but can be best explained by perturbations due to planetary companions. While the dynamical signature for a single planet is relatively simple, the dynamical signature for multiple planet systems can be much more complex and require detailed modelling. For multiple planet systems with significant planet-planet interactions, full n-body simulations are necessary to achieve self-consistent orbital solutions (Laughlin & Chambers 2001). When early observations are consistent with multiple qualitatively different orbital solutions (e.g., Goździewski & Konacki 2006), dynamical modelling can identify which epochs are particularly powerful for constraining models, resulting in increased efficiency of observations (Loredo & Chernoff 2003). Similarly, by assuming long-term dynamical stability, theorists can reject otherwise plausible orbital solutions and constrain the masses and orbital parameters for others (e.g., Rivera & Lissauer 2000).

3.1 Classes of multiple planet systems

The known multiple planet systems can be roughly classified into one of three categories:

1. hierarchical planetary systems that do not have significant orbital evolution,

2. secular planetary systems, those with significant secular orbital evolution, but without mean-motion resonances, and

3. mean-motion resonant planetary systems with significant mean motion resonances.

3.1.1 Hierarchical planetary systems

We mention the hierarchical systems only briefly. Since they show no significant interactions, these systems do not necessarily provide significantly more constraints than if they were two single planets. In practice, many of these systems may contain additional planets that have yet to be detected (Barnes & Raymond 2004). If such planets exist between the known planets, then they may provide for an increased coupling, in which case these systems may actually have significant dynamical interactions. On the other hand, dynamical studies of systems with significant dynamical interactions often reveal relatively few regions where an additional planet would be stable between the already known planets. In such cases, any additional planets are most likely to be found well inside or outside currently known planets.

3.1.2 Secular planetary systems

In secular planetary systems, the current orbital eccentricities of the planets will inevitably evolve under the mutual gravitation perturbations of the known planets. In a few cases, the orbital periods, mean longitudes of the planets, radial velocity amplitudes, eccentricities, and arguments of periastron are all precisely measured by a large data set of many high-precision radial velocities spanning multiple orbital periods of each planet (e.g., v And, GJ 876). In these cases, the secular evolution of the planetary system can be well characterized, with the greatest uncertainties being the unknown inclination of the orbital planes to the plane of the sky, the longitudes of ascending nodes, and the mass of the star. (In some systems with short-period planets, the stellar quadupole moment can also contribute to the uncertainty of long-term orbital integrations.) Numerical integrations can be used to identify inclinations (and hence planet masses) for which the mutual interactions would cause deviations from the observed radial velocities or dynamical instabilities would arise on timescales less than the system (presumed to be close to the age of the star). By rejecting the orbital solution likely to result in dynamical instabilities, theorists can place limits on the possible masses and orbital parameters (Rivera & Lissauer 2000). The secular evolution of the remaining models can then be evaluated, subject only to the caveat that additional low-mass and/or long-period planets might influence the dynamics. In strongly coupled systems, it may be possible to show that any undiscovered planets are unlikely to qualitatively change the secular orbital evolution.

Unfortunately, for many secularly evolving planetary systems, the observations are not yet able to place precise constraints on some important orbital parameters, such as eccentricity and argument of periastron. In some cases (e.g., systems where the orbital period of the outer planet is comparable or greater than the time span of high-precision observations), even the orbital period and radial velocity amplitude can be highly uncertain (Ford 2005).

3.1.3 Resonant planetary systems

In planetary systems containing pairs of planets near mean-motion resonances, the mutual planetary perturbations can have observable consequences over the time span of existing radial velocity surveys. In some cases, the perturbations cause deviations of the actual radial velocities from a model using a superposition of two Keplerian orbits. The best example is the system GJ 876 with two planets (b & c) in a 2:1 mean-motion resonance that results in the rapid precession of both planets (one full rotation every ∼9 years). In this case the precession is particularly rapid due to the short orbital period, low stellar mass, and relatively high planet masses (Laughlin & Chambers 2001). In most known multiple planet systems, the interactions are more subtle and/or have somewhat longer timescales than in GJ 876. Nevertheless, it can be practical to characterize properties such as the (anti-)alignment, rate of precession, and libration amplitudes of these systems by combining dynamical modeling with many high-precision radial velocity observations (Laughlin et al. 2005).

3.2 Multiple planet systems: probes of planet formation

Extrasolar planetary systems have the potential to provide constraints on models of planet formation and mechanisms for orbital migration and eccentricity evolution. Studying the orbital dynamics of multiple planet systems offer one particularly powerful tool for such research.

3.2.1 GJ 876 bc

The outer two giant planets around GJ 876 are in a 2:1 mean-motion resonance (Lee & Peale 2002). Given the small semi-major axes ($\sim$ 0.1 & 0.2 AU), it is unlikely that the planets formed in their present locations. Given the mean-motion resonance, it appears that the planets formed with a greater difference in their orbital semi-major axes and both migrated inwards towards the sun, but with the outer planet migrating slightly more rapidly. Such convergent migration can naturally result in resonant capture in a low-order resonance such as the 3:1 or 2:1 mean-motion resonance. Once the two planets are captured in resonance, continued migration will result in eccentricity excitation. The outer planet has only a small eccentricity, while the inner giant planet has an eccentricity of $\simeq$0.2. These values can be compared to those predicted by theoretical models of smooth migration in a disk leading to resonant capture. Lee & Peale (2002) found that the observed values are among the possible outcomes, but that they imply either: 1) the migration halted shortly after resonant capture, or 2) there was strong eccentricity damping during the migration. They dismiss the first option as unlikely, and focus on the second possibility, finding that the eccentricity damping timescale would need to be approximately two orders of magnitude shorter than the timescale for orbital decay. This provides one of the best observational constraints on the rate of eccentricity excitation during migration.

Since the planets' orbital periods are measured very accurately, the observed precession rate reduces to a complex function of the planet-star mass ratios, planet eccentricities, and orbital inclinations. Unfortunately, the eccentricities still have significant uncertainty, so there is a degeneracy between the eccentricities and inclinations. By using full n-body integrations (rather than a linear superposition of planets on non-interacting Keplerian orbits), the observations can constrain the planet masses and inclinations. Such an analysis suggests that the libration amplitudes are all relatively small (Laughlin et al. 2005). The Hubble Space Telescope's Fine Guidance Sensors have been used to make astrometric measurements of this system. Unfortunately, these observations have been interpreted as implying an inclination different than that of the dynamical analyses of the much larger radial velocity data set (Benedict et al. 2002). The reason for this difference is not yet clear.

3.2.2 v Andromedae bc

The outer two giant planets around v And are not in any mean-motion resonances, but do undergo significant eccentricity evolution on secular timescales.
Soon after their discovery, it was realized that mutual planetary perturbations
could cause significant secular evolution of the eccentricities and longitudes
of periastron for the outer two planets. Malhotra (2002) proposed a model in
which the outer planet was perturbed *impulsively*, as would be expected if it
had a close encounter with another (undetected) planet. In this model, the
periapses of the two planets could be either librating or circulating about an
aligned (or anti-aligned) configuration, depending on the relative phases at
the time of the impulsive perturbation. If the system were librating, then this
model would generally predict that the libration amplitude would be large and
that there would be significant eccentricity oscillations. An intensive radial velocity campaign has since been able to measure the current angle between the
two periapses to be $\Delta\omega \simeq 38° \pm 5°$. This implies a large libration amplitude
and supports models with an impulsive eccentricity perturbation. While the
eccentricity of the outer planet undergoes small oscillations, the eccentricity
of the middle planet undergoes very large oscillations with e ranging from
from 0.34 to very nearly zero. Ford, Lystad & Rasio (2005) used a rigorous
Bayesian statistical analysis of the radial velocity observations to determine
that the eccentricity of the middle planet periodically returns to nearly zero
for *all allowed* orbital solutions (see Figure 2 of Ford, Lystad & Rasio 2005).
This provides a strong constraint on the timescale for eccentricity excitation
in v And ($\simeq$ 100yr), thereby supporting the model of the outer planet being
perturbed impulsively by strong planet-planet scattering.

3.2.3 HD 128311 bc & HD 73526 bc

Other multiple planet systems likely offer additional insights into their orbital
histories and planet formation. For example, HD 128311 contains a pair of
planets near a 2:1 mean-motion resonance, again suggesting convergent migration leading to resonant capture. However, the eccentricity excitation in
this system might be quite unlike that of GJ 876. In particular, Sándor &
Kley (2006) proposed a hybrid scenario that invokes convergent migration,
resonant capture, and strong planet-planet scattering to explain the current
orbital dynamics of the best-fit orbital solution. While such a dynamical history is consistent with the published radial velocity data for HD 128311 and
HD 73526 (Sándor & Kley 2006; Sándor et al. 2007), the observations are
still consistent with a range of orbital solutions that is too broad to allow
a unique interpretation (e.g., Goździewski & Konacki 2006). Clearly, continued observations of these and other multiple planet systems is essential for
understanding the dynamical history of these systems.

4   Future tests of planet formation models

4.1   Radial velocity observations

Patiently extending existing radial velocity surveys to longer durations will enable the detection and orbital characterization of giant planets with orbital periods of over a decade (Wright et al. 2007). Such observations will address questions such as "How often do giant planets form and remain at orbital separations comparable to Jupiter?" and "Are most planets with masses and orbital periods comparable to Jupiter typically on circular or eccentric orbits?". Thus, long-term radial velocity observations could determine if the presently known planetary systems are typical or just the tip of an iceberg, with a higher frequency of giant planets at larger orbital separations. Similarly, if the excitation of orbital eccentricities were largely due to interactions during migration, then giant planets at large separations (i.e., with orbits beyond the ice line so that they presumably did not experience large scale migration) could have typical eccentricities smaller than those of the giant planets found inside of ∼2AU. In the other hand, if orbital eccentricities are typically excited by dynamical instabilities that propagate throughout a planetary system, then the eccentricity distribution may be insensitive to orbital separation.

As another example, deep radial velocity searches (i.e., many high-precision observations of a modest number of stars) will search for low-mass and likely rocky planets around nearby stars. These will provide valuable constraints on the frequency of terrestrial mass planets with short to intermediate orbital periods (days to ∼ 1 year). Intensified follow-up observations of the planets discovered by these surveys will be able to characterize their orbital eccentricities. "Do terrestrial mass planets typically have similar, smaller, or larger eccentricities than giant planets with similar orbital periods?" In some planetary systems, other eccentric giant planets, would inevitably excite the eccentricities of any detectable terrestrial planets. Therefore, it will be particularly interesting to characterize the eccentricity distribution of terrestrial mass planets in systems without nearby giant planets. Thus, we can determine if the dominant eccentricity excitation mechanisms are sensitive to planet mass.

Multiple planet systems can also benefit from intensive radial velocity campaigns. Remember that studies of the dynamical evolution of multiple planet systems are typically limited by the uncertainties in the planet masses and orbital parameters. Increasing the number of high-precision radial velocity observations improves the precision of orbital models and can eventually lead to deciphering the orbital history of these systems. Based on existing radial velocity surveys, finding each multiple planet system requires surveying ∼ 100 stars. Given the value of multiple planet systems for studying the dynamical evolution of planetary systems, it would be a shame not to allocate the follow-up resources necessary to get the most science return from these systems. Given the tendency for systems with one detected giant planet to have another detectable giant planet (Wright et al. 2006), intensive observations of multiple planet systems can simultaneously search for additional planets, including low-mass planets (e.g., GJ 876d, ρ Cnc e).

In addition to long and deep planet searches, complimentary wide planet searches (i.e., many target stars) are capable of discovering rare planetary systems that can provide clues to the limits of planet formation. Given the history of exoplanet discoveries, it seems likely that wide planet searches will continue to uncover surprising planetary systems. Additionally, wide surveys will enable more detailed statistical comparisons between observations and planet formation models (e.g., Ida & Lin 2000). Presently, the sample of known planets provides useful constraints on one-dimensional distributions (e.g., distribution of eccentricities or orbital periods), but is only beginning to enable tests of specific hypotheses that predict correlations in two dimensions (e.g., planet mass and orbital period, eccentricity and Safronov number; Ford & Rasio 2007). Performing systematic model-independent searches for correlations in planet masses and orbital properties will require significantly larger samples.

4.2   Transit searches

Recently, transit searches have become more efficient in detecting giant planets, including several with surprising sizes, orbital periods, and eccentricities. These offer the potential to significantly increase the sample of known short-period planets. With the launch of the CoRoT and soon Kepler space observatories, the transit method is soon expected to characterize the frequency and period distribution of terrestrial-sized planets.

Increasing the sample size of transiting planets will be particularly useful for testing planet formation models. For example, the inner edge of the orbital separations of short-period planets is sensitive to both the processes of orbital migration and the halting of migration. In particular, if the limits to short-period planets are set by the Roche limit, then the joint distribution of orbital separation, planet-star mass ratios, and planet radius can be used to recognize whether planets migrate while maintaining nearly circular orbits or whether they are often scattered inwards (Ford & Rasio 2006). If multiple mechanisms contribute to the population of short-period planets, then it will be particularly important to obtain a large sample size. For example, the more massive short-period planets (e.g., τ Boo b), the eccentric short-period planets (e.g., HAT-P-2b) or the very-short period planets ($P < 2\text{d}$; e.g., OGLE-TR-56b) might have a different dynamical history than the more common "hot-Jupiters" with orbital periods of $3 - 4$ days and typical masses of $0.5 M_J$.

Previous analyses assumed that the slope of the edge was set by the Roche limit and used the exoplanet population to determine the most probable location of the edge. Once planet searches uncover a sizable sample of short-period low-mass planets, it will be possible to test this assumption. Previous studies confined to giant planets were able to assume that the planet radius was insensitive to planet mass. Since the radius of terrestrial-mass planets is more sensitive to planet mass, transiting terrestrial planets (for which radii can be measured) will be particularly useful. The statistical analysis of ground-based transit surveys is complicated due to the complex selection effects, particularly with respect to orbital period (Gould et al. 2005). The new generation of space-based transit searches should find many transiting planets, and, more importantly, have more uniform and well-characterized selection effects, increasing the utility of statistical analyses.

Short-period transiting planets also enable exciting follow-up observations, particularly for planets around relatively bright stars ($V \leq 13$). For example, spectra taken during transit can study the planet atmosphere (Charbonneau et al. 2002) and measure the projection of the inclination of the planet's orbital angular momentum relative to the stellar rotation axis (Winn et al. 2005). Smooth migration in a gaseous disk predicts small inclinations. On the other hand, either planet scattering or perturbation from a binary companion followed by tidal circularization would predict a broad distribution of inclinations (Chatterjee et al. 2007; Fabrycky & Tremaine 2007). Early results for a couple of planets demonstrate that some systems are well aligned, but for other systems the existing measurement uncertainties leave room for significant improvement. Further observations will be needed to obtain precise measurements of the inclination angle for several additional systems in order to provide significant constraints on migration models.

The frequent smooth migration of planets is expected to result in the capture of additional planets into mean-motion resonances. Since the planet mass function appears to rise towards smaller masses, low-mass planets lurking in mean-motion resonances could be common. Unfortunately, detecting low-mass planets at the interior 2:1 mean-motion resonance with radial velocities is difficult due to a degeneracy with the orbital eccentricity of the outer planet. On the other hand, if the resonance is broken (e.g., due to the dissipation of the disk or a second stage of migration in a planetesimal disk), then radial velocity surveys should already be sensitive to such planets, if sufficiently massive. Fortunately, the transit timing technique is extremely sensitive to such planets either in or near the 2:1 (and other) mean-motion resonances. Even existing ground-based transit timing campaigns (e.g., Transit Light Curve project) have the sensitivity to detect sub-Earth-mass planets in resonance with a typical transiting hot-Jupiter (Agol et al. 2005; Holman & Murray 2005). Observational campaigns have only begun to make such measurements and many transit times will be required before it is possible to recognize to transit timing signature of such planets (Ford & Holman 2007).

4.3 Long-term

Clearly, much remains to be learned about the formation and dynamical evolution of planetary systems. The field is simultaneously wide open for new theoretical advances and in desperate need of more observational constraints. So far, radial velocity surveys have provided the overwhelming majority of exoplanet discoveries. In the coming years, radial velocity observations will continue to play an important role in providing observational constraints for testing theories for the formation and dynamical evolution of planetary systems. The rapidly accelerating detection rate of ground-based and space-based transit searches is expected to significantly extend the number and mass range of known extra-solar planets. In the longer term, other planet detection methods (e.g., microlensing, transit timing, direct detection, astrometry) can also be expected to provide new and complimentary windows for exploring extrasolar planets. Together, these will provide the observational data that will form the basis for revising models of planet formation for the foreseeable future.

Acknowledgments

I thank the sponsors and organizers of the summer school, particularly Bonnie Steves and Martin Hendry for their hospitality and fostering such a stimulating environment. I acknowledge the helpful suggestions for Fred Adams, Phil Armitage, Scott Gaudi, Bo Gustafson, Willy Kley, Renu Malhotra, Mathew Holman, Geoff Marcy, Frederic Rasio, Zolt Sandor, Steinn Sigurdsson. Support for EBF was provided in part by NASA through Hubble Fellowship grant HST-HF-01195.01A awarded by the Space Telescope Science Institute, which is operated by the Association of Universities for Research in Astronomy, Inc., for NASA, under contract NAS 5-26555.

References

Adams, F. & Laughlin, G. (2003), 'Migration and dynamical relaxation in crowded systems of giant planets', *Icarus* **163**, 290–306.

Adams, F. & Laughlin, G. (2006), 'Effects of secular interactions in extra-solar planetary systems', *Astrophysical Journal* **649**, 992–1003.

Adams, F. et al. (2006), 'Early evolution of stellar groups and clusters: environmental effects on forming planetary systems', *Astrophysical Journal* **641**, 504–525.

Agol, E. et al. (2005), 'On detecting terrestrial planets with timing of giant planet transits', *Monthly Notices of the RAS* **359**, 567–579.

Andrews, S. & Williams, J. (2007), 'A submillimeter view of protoplanetary dust disks', *Astrophysics and Space Science* pp. 363–+.

Armitage, P. & Hansen, B. (1999), 'Early planet formation as a trigger for further planet formation', *Nature* **402**, 633–635.

Armitage, P. et al. (2002), 'Predictions for the frequency and orbital radii of massive extra-solar planets', MNRAS **334**, 248–256.

Artymowicz, P. (1992), 'Dynamics of binary and planetary-system interaction with disks - eccentricity changes', PASP **104**, 769–774.

Barnes, R. & Raymond, S. (2004), 'Predicting planets in known extra-solar planetary systems. I. Test particle simulations', *Astrophysical Journal* **617**, 569–574.

Barranco, J. & Marcus, P. (2005), 'Three-dimensional vortices in stratified protoplanetary disks', *Astrophysical Journal* **623**, 1157–1170.

Benedict, G. et al. (2002), 'A mass for the extra-solar planet Gliese 876b determined from Hubble space telescope fine guidance sensor 3 astrometry and high-precision radial velocities', *Astrophysical Journal, Letters* **581**, L115–L118.

Bromley, B. & Kenyon, S. (2006), 'A hybrid N-body-coagulation code for planet formation', *Astronomical Journal* **131**, 2737–2748.

Butler, R. et al. (2006), 'Catalog of nearby exoplanets', *Astrophysical Journal* **646**, 505–522.

Chambers, J. (2006), 'Planet formation with migration', *Astrophysical Journal, Letters* **652**, L133–L136.

Chambers, J., Wetherill, G. & Boss, A. (1996), 'The stability of multi-planet systems', *Icarus* **119**, 261–268.

Charbonneau, D. et al. (2002), 'Detection of an extra-solar planet atmosphere', *Astrophysical Journal* **568**, 377–384.

Chatterjee, S., Ford, E. & Rasio, F. (2007), 'Dynamical outcomes of planet-planet scattering', *ArXiv Astrophysics e-prints*.

Dobbs-Dixon, I., Li, S. & Lin, D. (2007), 'Tidal barrier and the asymptotic aass of proto-gas giant planets', *Astrophysical Journal* **660**, 791–806.

Eisner, J. et al. (2007), 'Near-infrared interferometric, spectroscopic, and photometric monitoring of T Tauri inner disks', *ArXiv e-prints* **707**.

Fabrycky, D. & Tremaine, S. (2007), 'Shrinking binary and planetary orbits by Kozai cycles with tidal friction', *ArXiv e-prints* **705**.

Ford, E. (2005), 'Quantifying the uncertainty in the orbits of extra-solar planets', *Astronomical Journal* **129**, 1706–1717.

Ford, E. & Chiang, E. (2007), 'The formation of ice giants in a packed oligarchy: instability and aftermath', *Astrophysical Journal* **661**, 602–615.

Ford, E., Havlickova, M. & Rasio, F. (2001), 'Dynamical instabilities in extra-solar planetary systems containing two giant planets', *Icarus* **150**, 303–313.

Ford, E. & Holman, M. (2007), 'Using transit timing observations to search for Trojans of transiting extra-solar planets', *ArXiv e-prints* **705**.

Ford, E., Kozinsky, B. & Rasio, F. (2000), 'Secular evolution of hierarchical triple star systems', *Astrophysical Journal* **535**, 385–401.

Ford, E., Lystad, V. & Rasio, F. (2005), 'Planet-planet scattering in the upsilon Andromedae system', *Nature* **434**, 873–876.

Ford, E. & Rasio, F. (2006), 'On the Relation between hot Jupiters and the Roche limit', *Astrophysical Journal, Letters* **638**, L45–L48.

Ford, E., Rasio, F. & Sills, A. (1999), 'Structure and evolution of nearby stars with planets. I. Short-period systems', *Astrophysical Journal* **514**, 411–429.

Goldreich, P., Lithwick, Y. & Sari, R. (2004), 'Planet formation by coagulation: A focus on Uranus and Neptune', *Annual Review of Astron. and Astrophys.* **42**, 549–601.

Goldreich, P. & Sari, R. (2003), 'Eccentricity evolution for planets in gaseous disks', *Astrophysical Journal* **585**, 1024–1037.

Goldreich, P. & Tremaine, S. (1980), 'Disk-satellite interactions', *Astrophysical Journal* **241**, 425–441.

Goldreich, P. & Ward, W. (1973), 'The formation of planetesimals', *Astrophysical Journal* **183**, 1051–1062.

Gould, A. et al. (2006), 'Frequency of hot Jupiters and very hot Jupiters from the OGLE-III transit surveys toward the galactic bulge and carina', *Acta Astronomica* **56**, 1–50.

Goździewski, K. & Konacki, M. (2006), 'Trojan pairs in the HD 128311 and HD 82943 planetary systems?', *Astrophysical Journal* **647**, 573–586.

Hahn, J. & Malhotra, R. (2005), 'Neptune's migration into a stirred-Up Kuiper belt: A detailed comparison of simulations to observations', *Astronomical Journal* **130**, 2392–2414.

Holman, M. & Murray, N. (2005), 'The use of transit timing to detect terrestrial-mass extra-solar planets', *Science* **307**, 1288–1291.

Holman, M., Touma, J. & Tremaine, S. (1997), 'Chaotic variations in the eccentricity of the planet orbiting 16 Cygni B', *Nature* **386**, 254–256.

Holman, M. et al. (2006), 'The transit light curve project. I. Four consecutive transits of the exoplanet XO-1b', *Astrophysical Journal* **652**, 1715–1723.

Ida, S. & Lin, D. (2004), 'Toward a deterministic model of planetary formation. I. A desert in the mass and semimajor axis distributions of extra-solar planets', *Astrophysical Journal* **604**, 388–413.

Juric, M. & Tremaine, S. (2007), 'The eccentricity distribution of extra-solar planets', *ArXiv Astrophysics e-prints*.

Kokubo, E. & Ida, S. (2002), 'Formation of protoplanet systems and diversity of planetary systems', *Astrophysical Journal* **581**, 666–680.

Laughlin, G. & Adams, F. (1998), 'The modification of planetary orbits in dense open clusters', *Astrophysical Journal, Letters* **508**, L171–L174.

Laughlin, G. & Chambers, J. (2001), 'Short-term dynamical interactions among extra-solar planets', *Astrophysical Journal, Letters* **551**, L109–L113.

Laughlin, G. et al. (2005), 'The GJ 876 planetary system: A progress report', *Astrophysical Journal* **622**, 1182–1190.

Lee, M. & Peale, S. (2002), 'Dynamics and origin of the 2:1 orbital resonances of the GJ 876 planets', *Astrophysical Journal* **567**, 596–609.

Levison, H. & Morbidelli, A. (2007), 'Models of the collisional damping scenario for ice-giant planets and Kuiper belt formation', *Icarus* **189**, 196–212.

Lin, D. & Ida, S. (1997), 'On the origin of massive eccentric planets', *Astrophysical Journal* **477**, 781–+.

Malhotra, R. (2002), 'A dynamical mechanism for establishing apsidal resonance', *Astrophysical Journal, Letters* **575**, L33–L36.

Meyer, M. et al. (2006), 'The formation and evolution of planetary systems:
placing our solar system in context with Spitzer', PASP **118**, 1690–1710.

(2005), 'On the origin of the eccentricities of extra-solar planets', *Astronomical
Journal* **130**, 280–294.

Nelson, R. & Papaloizou, J. (2002), 'Possible commensurabilities among pairs
of extra-solar planets', MNRAS **333**, L26–L30.

Papaloizou, J. & Szuszkiewicz, E. (2005), 'On the migration-induced resonances in a system of two planets with masses in the Earth mass range',
MNRAS **363**, 153–176.

Papaloizou, J. & Terquem, C. (2001), 'Dynamical relaxation and massive
extra-solar planets', MNRAS **325**, 221–230.

Rafikov, R. (2005), 'Can giant planets form by direct gravitational instability?', *Astrophysical Journal, Letters* **621**, L69–L72.

Rivera, E. & Lissauer, J. (2000), 'Stability analysis of the planetary system
orbiting v Andromedae', *Astrophysical Journal* **530**, 454–463.

Sándor, Z. & Kley, W. (2006), 'On the evolution of the resonant planetary
system HD 128311', *Astronomy and Astrophysics* **451**, L31–L34.

Sándor, Z., Kley, W. & Klagyivik, P. (2007), 'Stability and formation of the
resonant system HD 73526', *Astronomy and Astrophysics* **472**, 981–992.

Takeda, G. & Rasio, F. (2005), 'High orbital eccentricities of extra-solar planets induced by the Kozai mechanism', *Astrophysical Journal* **627**, 1001–
1010.

Terquem, C. & Papaloizou, J. (2007), 'Migration and the formation of systems
of hot super-earths and Neptunes', *Astrophysical Journal* **654**, 1110–1120.

Thommes, E., Duncan, M. & Levison, H. (1999), 'The formation of Uranus
and Neptune in the Jupiter-Saturn region of the solar system', *Nature*
402, 635–638.

Trilling, D., Lunine, J. & Benz, W. (2002), 'Orbital migration and the frequency of giant planet formation', *Astronomy and Astrophysics* **394**, 241–
251.

Tsiganis, K. et al. (2005), 'Origin of the orbital architecture of the giant
planets of the Solar System', *Nature* **435**, 459–461.

(1997), 'Survival of planetary systems', *Astrophysical Journal, Letters* **482**, L211+.

Winn, J. et al. (2005), 'Measurement of spin-orbit alignment in an extra-solar
planetary system', *Astrophysical Journal* **631**, 1215–1226.

Wright, J. et al. (2007), 'Four new exoplanets and hints of additional substellar
companions to exoplanet host stars', *Astrophysical Journal* **057**, 533–545.

Zakamska, N. & Tremaine, S. (2004), 'Excitation and propagation of eccentricity disturbances in planetary systems', *Astronomical Journal* **128**, 869–
877.

Late stages of solar system formation and implications for extra-solar systems

Kleomenis Tsiganis

Section of Astrophysics, Astronomy & Mechanics, Department of Physics
Aristotle University of Thessaloniki, GR 54 124, Thessaloniki, Greece

In this chapter we will discuss our current understanding of the late stages of formation of the solar system. We shall focus mainly on dynamical processes that took place after the giant planets (Jupiter, Saturn, Uranus and Neptune) had already formed. We will try to explain how the interaction of the giant planets with the remaining protoplanetary disc has led our planetary system to develop the currently observed dynamical architecture. More precisely, we will discuss the two main phases of *planet migration*, i.e. (i) gas-driven migration, when enough gas was still present in the system, and (ii) planetesimals-driven migration, that started immediately after the gas was gone. We will show that, during the first migration phase, the outer planets could have reached a stationary, multi-resonant, state, the outermost one being only $\sim 12 - 13$ AU away from the Sun. Then, during the phase of planetesimals-driven migration, the planets could have had their eccentricities increased by resonance-crossing, a mechanism that leads to a short instability phase and a complete reshaping of the planetary orbits. We will show that this model can explain a number of puzzling problems, such as the orbital parameters of the giant planets, the orbital distribution in the small bodies' reservoirs and the occurrence of the Late Heavy Bombardment (LHB) of the terrestrial planets. Finally, we will examine the aforementioned dynamical mechanisms in the more general context of Extra-solar planetary systems and we will attempt to develop a general evolution scheme for young planetary systems.

1 Formation stages of planetary systems

Planets are born in protoplanetary discs, which are formed around young stars and are composed of gas (primarily hydrogen and helium) and a small amount of solids, in the form of dust. The formation of dust grains and the exact mechanisms by which planets are later formed are outside the purpose of this chapter. For a review see Armitage (2007), Papaloizou & Terquem (2006) and Chapter 7 by E. Ford in this book. Here, we will only briefly outline the main phases of giant planet formation. Note also that, throughout this chapter, we will focus on the evolution of the giant planets of the solar system and not discuss the formation and evolution of terrestrial planets. It is generally believed that terrestrial planets are formed by mutual collisions and coagulation of planetary embryos during the very last evolutionary stages of the planetary system, as the system evolves towards a "mature" state in which the masses and orbits of the giant planets no longer change. As terrestrial planets' formation cannot really affect the evolution of the giant planets, this process will not be considered.

The sedimentation of dust grains in the mid-plane of the gaseous disc marks the onset of planet formation. Coagulation of grains leads to the formation of small rocks, that gradually evolve to km-sized bodies, called *planetesimals* - a process whose details are not well understood thus far. These are the building blocks of the planetary cores. Mutual collisions lead to the formation of lunar-sized embryos, via a process known as a runaway growth. The solid cores of the giant planets form by collisions and merging of such embryos. The planetary cores settle on orbits that are separated enough for the system to be stable (Kokubo & Ida 2000). We note that core formation requires a large amount of solids in the form of dust grains. This condition is better satisfied if water is in the form of ice, instead of vapour. As the temperature of the disc is a decreasing function of heliocentric distance, this condition is met only a few AU away from the star, beyond a limiting distance called the *snow line*. In the case of the solar system, the snow line is located at $\sim 3 - 5$ AU. Thus, in the solar system, it would be much easier to form the core of a giant planet beyond $\sim 3 - 5$ AU, where the giant planets are indeed located. Similarly, the inner parts of the solar system would have formed "dry" planetesimals, as is the case with inner-belt asteroids.

The subsequent growth of the planet to a gas giant is still a matter of debate. Pollack et al. (1996) assume that planetary cores of $10 - 15$ Earth masses can accrete gas such that an envelope is slowly formed, which subsequently collapses on the core. On the other hand, Boss (1997, 2000) proposes that gas giants can form rapidly, by gravitational instability. Independent of which of the two models better explain the solar system, at the end, a number of gas/ice giants had to form in the protoplanetary disc. The formation times range from a few 10^5 yrs up to a few 10^6 yrs, depending on the model. As protoplanetary discs have a physical lifetime of ~ 10 My, the newly formed planets had to interact with the remaining, still massive, gaseous disc. In the following we will assume the four outer planets of the solar system to have reached their final masses (no accretion), and to be embedded in a massive gaseous disc of ~ 0.01 $M_\odot$. The subsequent evolution of the planets will be dictated by their dynamical interaction with the disc and in particular by the exchange of angular momentum between the planets and the disc. This interaction leads to *planet migration,* i.e. a large-scale variation of the semi-major axes of the planetary orbits. Migration will be reviewed in the following section.

A few Myrs after the formation of the planets, the gaseous disc is no longer present, owing to viscous accretion, stellar winds and photo-evaporation by the host star. At that point the planetary system consists of the giant planets (probably in different orbits than their original ones) and a remnant disc of planetesimals (the leftovers of core formation) in which the planets are embedded. Recent theoretical works on the solar system suggest that this disc had a mass of $\sim$ 50 Earth masses and extended to $\sim$ 30 − 35 AU away from the Sun (see Section 2). In other systems the disc may extend to even greater distances. The exchange of angular momentum between the planets and the planetesimals' disc results in a second phase of planet migration. This process may or may not be 'smooth,' as possible resonant interactions between the planets may increase their orbital eccentricities and lead to planet-planet scattering. As we will see in Section 3, accurate modelling of this process can explain a great number of puzzling observations in the solar system.

After the planets destroy the planetesimals' disc, planet migration stops. When this stage is reached, the system is entering its maturity. There is practically no more mass for the planets to either accrete or disperse away. Thus, the planets settle on their final orbits. The time interval between planet formation and the end of migration is between a few tens to several hundreds of My – depending on the characteristics of each system. For the solar system, we believe that it took $\sim$ 800 My, i.e. a small fraction of its age (4.6 Gy). Thus, the solar system has had the same dynamical structure for the last $\sim$ 3.8 Gy, with the exception of the small body populations. However, as we will see in the following, the main characteristics of the orbital distribution of even the small-body populations were shaped during the late stages of planet migration.

2   Planet migration

In this section we will briefly review the two main mechanisms that lead to planet migration in general planetary systems: (i) gas-driven migration, which takes place during the first few My after the formation of the giant planets, and (ii) planetesimals-driven migration, which effectively starts after the dissipation of the gas disc and may last for several hundreds of My. Although the case of the solar system will be treated more thoroughly in Section 3, we will point out in this section a few relevant issues.

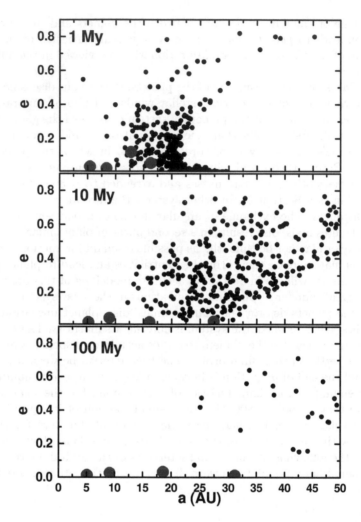

Figure 1. *Evolution of the semi-major axes of the outer planets (largest circles), in a simulation of planetesimals-driven migration (planetesimal - smallest circles). Time is shown on the top-left corner of each of the three snapshots. The planets move outwards (except for Jupiter), reaching final orbits that are similar to their currently observed ones (in terms of a).*

2.1   Gas-driven migration

The main focus of this chapter will be on planetesimals-driven migration. For this reason we will limit our discussion on gas-driven migration to those points that are necessary for the reader to understand the course of events in solar system history. For a more complete description we refer the reader to the reviews by Armitage (2007), Papaloizou & Terquem (2006) and Papaloizou et al. (2007).

Consider a planet of mass m_p, orbiting a star of mass M_*, surrounded by a protoplanetary gaseous disc of mass m_d. The planet and the disc interact gravitationally, perturbing each other and exchanging angular momentum. At the same time, the disc itself evolves under the effects of viscous accretion, self-gravity, photo-evaporation, etc. The evolution of discs under different conditions is described in Armitage (2007). Here we consider a hydrodynamical disc, whose evolution is dominated by viscous accretion onto the central star.

The embedded planet perturbs the disc, by exciting waves. This leads to the formation of a *spiral wake*, a structure which revolves around the star at the same angular velocity as the planet. The inner part of the disc, which revolves faster than the planet, is decelerated by the wake and gives angular momentum to the planet. Thus, the planet tends to move outwards and the inner disc tends to move inwards. The opposite is true for the outer, slower revolving disc: the planet gives angular momentum to the gas and tends to move inwards, while the disc moves outwards. The balance of angular momentum transport determines the direction and speed of planet migration.

In the linear approximation, it can be shown (Goldreich & Tremaine 1979, 1980) that the exchange of angular momentum between the disc and the planet averages to zero, except from the vicinity of *Lindblad resonances*, where

$$m(\Omega - \Omega_p) = \pm\kappa = \pm\Omega \tag{1}$$

and κ is the epicyclic frequency, which is equal to the revolution frequency Ω for Keplerian discs. Thus, the net flux of angular momentum is determined by the sum of the magnitude of the torques that are exerted on both the *inner* and *outer* Lindblad resonances.

When the planet's mass is small (say, less than 10 $M_\oplus$) the differential Lindblad torque is negative, meaning that the planet migrates inwards, at a rate that is inversely proportional to the planet's mass. This is what we refer to as *Type I* migration (Ward 1997). Tanaka et al. (2002) calculated the Type I migration rate for a 3-D isothermal disc with surface density profile $\Sigma(r) \sim r^{-\gamma}$,

$$\tau_I = -\frac{r_p}{\dot{r}_p} \sim \frac{M_*^2}{m_p \Omega_p \Sigma(r_p) r_p^2} \left(\frac{c_s}{r_p \Omega_p}\right)^2 \tag{2}$$

where c_s is the sound speed. For a planetary core of 5 $M_\oplus$, initially at 5 AU, the time needed to migrate all the way to the star is only a few times 10^5 y. This is a very rapid phenomenon that poses the problem of "delaying" Type I migration enough, so that the cores of the giant planets do not fall onto the star, before they become giants. Several possible solutions to this problem have been proposed, the most efficient being probably the "planet trap" mechanism, proposed by Masset et al. (2006). This effect is created by small density "jumps" in $\Sigma(r)$, around which the differential Lindblad torque can diminish to zero, thus halting Type I migration (see also Morbidelli et al. 2008).

It is reasonable to assume that, for the cores of the solar system giant planets, Type I migration was somehow halted, giving them time to grow. However, gas giants also migrate, but in a different way. Planets with $m_p \gg 10\, M_\oplus$ can create a "gap", repelling gas from the vicinity of their orbit. The conditions for creating a gap are related to (i) the ratio between the scale height of the disc, H, and the size of the planet's Hill sphere, and (ii) the rate at which viscosity and pressure effects try to replenish the gap (see Crida et al. 2006 for details). The formation of the gap diminishes the effects of Lindblad resonances; the planet essentially decouples from the distant disc. Hence, while the planet poses a tidal barrier to the material of the disc, it is forced to follow the viscous evolution of the disc, undergoing *Type II* migration, at a rate

$$\tau_{II} = \frac{2}{3\alpha\Omega}\left(\frac{h}{r}\right)_p^{-2} \tag{3}$$

where α is a dimensionless parameter characteristic of the viscosity profile in the disc (Shakura & Syunyaev 1973). This calculation assumes that the mass of the disc is much larger than m_p, in the vicinity of the planet. This is, however, not true for Jupiter-sized planets and more refined calculations show that the migration time actually decreases for higher values of m_p (Syer & Clarke 1995). For a Jupiter-sized planet initially at 5 AU and typical disc parameters, the time needed to reach the star is $\tau_{II} \sim 0.5 - 1$ My. Thus, Type II migration is certainly slower than Type I, but still too fast, compared to the typical lifetime of the disc (a few My). However, additional nonlinear phenomena may reduce the speed of migration such that, at least for "hot Jupiters", the lifetime of the disc becomes comparable to the inverse of the migration rate. Clearly, in our solar system, Type II migration was not that effective for Jupiter.

A third mode of migration, called Type III or *runaway migration* (Masset & Papaloizou 2003), is also possible, mainly for Saturn-sized planets. The onset of this migration mode depends on the amount of mass that is trapped in the corotation zone of the planet, relative to its own mass. This trapped material exerts a torque on the planet, which can give positive feedback to the planet's migration rate. We should note that all the aforementioned phenomena, in particular Type I migration, can be very different if we consider magnetized discs, where small planets can *random walk* inside a turbulent disc (Papaloizou et al. 2007).

The above description of gas-driven migration was limited to the interaction of a single planet with the disc. If two (or more) giant planets exist, the evolution can be very different. Differential migration of the planets can force them to get "locked" in an orbital (or, mean motion) resonance (hereafter MMR); this is how we believe that extra-solar systems that are observed in MMRs (especially with a 2:1 ratio of the orbital periods) have formed.

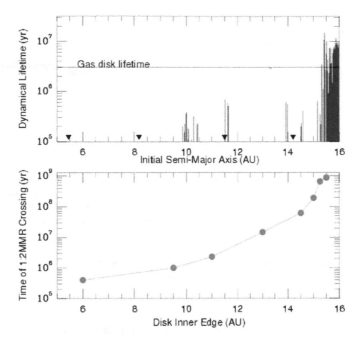

Figure 2. *(top) Dynamical lifetime of test particles as a function of ini-tal a, for the selected planetary configuration. As shown in the figure, only particles that are ∼ 1.5 AU away from Neptune can survive longer than the gas disc, thus driving migration. (bottom) Calculation of the instability time (resonance-crossing epoch), as a function of location of the inner edge of the disc. For the configuration implied by the results shown in the top panel, the time of instability is of the same order of magnitude as the estimated LHB delay. Reproduced from the results of Gomes et al. (2005).*

Resonance trapping generally increases the orbital eccentricities and can also increase the relative inclination of their orbital planes (Thommes & Lissauer, 2003). What is also interesting is that, after the planets are trapped in res-onance, they can continue their migration towards the star at a slower rate than before, while maintaining the commensurability. The values of the plan-etary masses seem to be critical, not only for determining the MMR at which the planets will be trapped (typically either the 1:2 or the 2:3 MMR), but also for the subsequent evolution of the resonant pair.

In Morbidelli & Crida (2007) it was shown that, for mass ratios similar to the Jupiter-Saturn pair, the small (exterior) planet tends to "jump" over the 1:2 MMR and get trapped in the 2:3 MMR. On the contrary, for heavier planets and pairs with mass ratio ∼ 1, capture in the 1:2 MMR is observed. Moreover, for cases similar to the Jupiter-Saturn pair and for a typical range of values for H and α, Type II migration of the pair can be halted, so that the

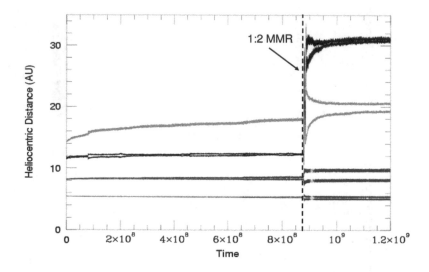

Figure 3. *Time evolution of the pericentric and apocentric distance of each planet, in a typical Nice-model simulation. The epoch of resonance crossing is marked on the graph. See text for a detailed description. Reproduced from the results of Gomes et al. (2005).*

resonant pair is "parked" at relatively large distances from the star. On the contrary, for mass ratios > 1 (i.e. the outer planet being more massive than the inner one), more extreme values for H and α should be adopted to halt - if possible - Type II migration. In other words, planet pairs of the Jupiter-Saturn type can avoid approaching the star, also maintaining a "tighter" resonance (2:3 instead of 1:2), while pairs of planets of similar (Jupiter-like) masses, as well as pairs with a heavier planet on the outside, tend to approach the star. These results are also supported by the hydrodynamical simulations of Pierens & Nelson (2008). A look in the exoplanets' catalogue confirms that two-planet resonant systems follow the above trends.

If more than two planets are considered, a multi-resonant state can be achieved. Morbidelli et al. (2007) have shown that this is a quite probable scenario for the outer planets of our solar system. In particular, in their simulations it was found that Saturn ends-up in a 3:2 MMR with Jupiter, Uranus in a 4:3 or 3:2 MMR with Saturn, and Neptune in a 5:4, or 4:3 MMR with Uranus. Moreover, this quadruple resonant state resists Type II migration, such that the system remains more or less stationary, until the disc disperses. This type of evolution is compatible with the existence and observed dynamical structure of the asteroid belt.

The above results allow us to imagine a scenario for the evolution of the giant planets in our solar system. In hydrodynamical simulations, it is found that mass is concentrated outside the gap that a massive planet opens in the disc. Thus, we can imagine that, if Jupiter was already formed, Saturn would form outside Jupiter's gap, as large planetesimals could pile up there, undergoing Type I migration, eventually being halted by the planet-trap mechanism of Masset et al. (2006). After the formation of Saturn, the planets would get locked in the 3:2 MMR and Type II migration would be halted. Similarly, Uranus and Neptune could be formed outside the gap of Saturn and Uranus respectively, and subsequently evolve into a MMR with one another. Thus, the system could end-up in a multi-resonant and nearly stationary state, which could remain stable long after the gas disc was gone. Of course this implies that the extent of the planetary system would be, at those times, much smaller than nowadays observed, the orbital radii of the planets being between 5 and 14 AU. Hence, planetesimals-driven migration had to occur subsequently, in order to explain the currently observed dynamical state of the solar system.

2.2 Planetesimals-driven migration

When the gas disc dissipates, planet migration stops. The planetary system, however, does not only contain the major planets. Outside the orbit of the outermost planet exists a relatively massive disc of planetesimals, which did not have enough time to form massive planetary cores. The linear dimensions of the disc can exceed several tens of AUs, while its mass can be several tens of Earth masses. In the solar system for example, in the beginning of the gas-free era, the remnant disc of planetesimals extended up to $\sim 30 - 35$ AU (Gomes et al. 2004) and had a mass of ~ 50 Earth masses. Fernandez & Ip (1984) were the first to point out that the exchange of angular momentum between the planets and the disc particles would force the planets to migrate. This pioneering work was not much appreciated, until the discovery of the Kuiper Belt by Jewitt and Luu (1993). Soon after the discovery of the high-eccentricity *plutinos* – objects trapped in a 2:3 orbital resonance with Neptune – Malhotra (1995) realized that the orbital distribution of these objects could not be understood, unless Neptune had migrated by an amount of at least 7 AU.

Schematically, planetesimals-driven migration proceeds as follows: when a disc particle approaches the outermost planet, it receives a "kick" that decreases (or increases) the semi-major axis of its orbit. Consequently, the planet must move outwards (or inwards) by a very small amount. Repeated encounters between the two bodies give zero net displacement, unless the kick is so strong that the particle is ejected from the system on a hyperbolic orbit, in which case the "symmetry" is broken and the planet moves inwards. If a steady flux of planetesimals to planet-crossing orbits exists, the balance between "sources" and "sinks" determines the direction and rate of migration. Simple analytic estimates exist (e.g. Ida et al. 2000), although numerical simulations, such as those of Gomes et al. (2004), reveal a much larger complexity of the process. For more details on planet migration in planetesimal discs, we refer the reader to the recent review by Levison et al. (2007).

When two (or more) planets exist in the system, the situation is more complex. A particle, deflected inwards by the outermost planet, can encounter one of the inner planets. In that case, the particle escapes the influence of the outer planet, whose orbital radius is slightly increased, and it subsequently evolves under the dominant influence of the inner planet (see Levison et al. 2007). The above scheme can be generalized to the case of more than two planets. In the case of the four giant planets of the solar system, the innermost planet (Jupiter) is far more massive than the outermost one. Numerical simulations show that most disc particles approach Jupiter, after being "kicked" inwards by Neptune, Uranus and Saturn, which ejects them on hyperbolic orbits. Consequently, Neptune, Uranus and Saturn move outwards, while Jupiter moves inwards. Of course, the more massive planets (Jupiter and Saturn) are displaced much less than the ice giants; the orbital radii of the first two change by a few tenths of an AU, while the ones of the ice giants may change by several AU. This is because, as we already saw, the displacement is inversely proportional to the mass of the scattering planet.

A numerical simulation of this process is shown in Figure 1. Three snapshots of the evolution of the system are shown, in which the points (largest circles) denote the four giant planets and the dots (smallest circles) are disc particles. In this simulation a disc of 35 Earth masses was used, extending up to 30 AU. The planets were started within 20 AU from the Sun, in accordance with the work of Malhotra (1995). As shown in the plots, the planets deplete the disc within 100 My and reach stable orbits with approximately the same values of semi-major axis as observed today. The remnants of the disc have an orbital distribution that is similar to the one of the current trans-Neptunian population. Indeed, as first shown by Malhotra (1995), planetesimals-driven migration can explain the final semi-major axes of the planetary orbits and the orbits of the plutinos. However, as we will see in the following section, this "smooth" migration model cannot explain important observations in the solar system.

3 Probing the history of the solar system

In studying the evolution of the solar system, we have to figure out what happened by looking at the final outcome (what we observe today). However, having a rough idea about the processes that took place, we can try to find the "good" range of parameters that reproduce as many of the observations as possible (and do not contradict any, if possible...). Thus, if we believe that planet migration took place, we have to look for (i) the initial conditions of the planets and (ii) the parameters of the planetesimals disc that provide satisfying explanations to (at least) the following observational facts:

1. The orbits of the outer planets.

2. The mass and orbital distribution of the asteroid belt, the Kuiper belt and other, special, small-body populations (Trojan asteroids, irregular satellites of the planets, etc.).

3. The Late Heavy Bombardment of the terrestrial planets and the Moon.

For point (1) above, several questions need to be answered. For example, how did Neptune reach 30 AU and why did it stop there? Why did Jupiter not move, as indicated by the orbital structure of asteroids in the outer belt (which would be decimated if Jupiter were displaced inwards by ~ 0.5 AU)? More important, if planet formation theories are correct and planets form on circular and co-planar orbits, how did the giant planets reach their currently observed eccentricities (up to 0.1 for Saturn) and mutual inclinations ($2 - 3$ deg)?

For point (2), one has to remember the following: The minimum mass solar nebula (MMSN) model suggests that the asteroid belt was at least 1,000 times more massive than observed today. How did the depletion take place? According to Petit et al. (2002) the formation of the terrestrial planets, by merging of planetary embryos that interacted gravitationally and "stirred-up" the asteroids, could deplete the belt by a factor of ~ 100. Thus, another factor of ~ 10 is still missing. A similar question has to do with the small estimated mass of the Kuiper Belt. Finally, a careful study of other, special populations of small bodies, such as Jupiter's Trojan asteroids, clearly shows that their orbital distribution cannot be explained by in situ formation and subsequent dynamical evolution in the framework of a "static" planetary system. We remind the reader that Trojans are asteroids moving in the vicinity of the stable Lagrangian points of the restricted Sun-Jupiter-asteroid problem, leading or trailing the planet by ±60 degrees in mean longitude.

Point (3) is certainly more subtle. The surface of the Moon is covered with large, lava-filled, basins and numerous small craters, which are the result of an intense bombardment of the Moon by small bodies. When the first lunar samples were analyzed, things became more complex. The values of the sample ages of the large basins formed a tight clustering around 3.9 Gy, suggesting a heavy bombardment of the Moon ~ 600 My after its formation, with a duration of only a few tens of Mys: this is referred to as the *Late Heavy Bombardment* (LHB, see Tera et al. 1974, Dalrymple and Ryder 1993, Cohen et al. 2000, and references in Gomes et al. 2005). An alternative scenario supports the continuously declining bombardment of the Moon for ~ 700 My, corresponding to the "tail" of terrestrial planet accretion. Several theoretical models were proposed to explain the LHB, but all of them turned out to violate some important observational constraint (see related references in Gomes et al. 2005). Of course the LHB must have had dramatic consequences for all objects in the inner solar system, including the Earth. However, its results are most clearly visible on the surface of the Moon, which does not suffer from plate tectonics, atmospheric phenomena, etc.

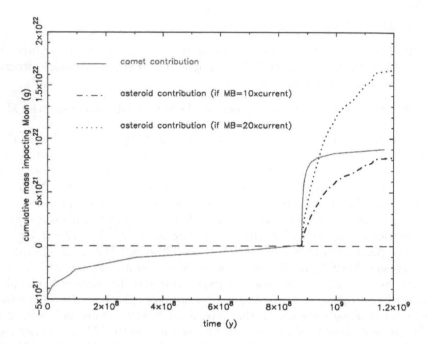

Figure 4. *Time evolution of the cumulative mass of projectiles hitting the Moon, in the simulation of Figure (3). The vertical axis is scaled to the epoch of resonance crossing. Planetesimals coming from the outer disc are labeled "comets". The contribution of "asteroids" is also measured, assuming two different values of the initial mass of the asteroid belt. Reproduced from the results of Gomes et al. (2005).*

From the dynamical point of view, the "declining bombardment" model, although easier to understand, does not seem probable. If we take into account the depletion of the asteroid belt during terrestrial planet formation, we find that there are not enough projectiles to sustain the necessary mass flux on the Moon for 600 My. Even more, recent results by Bottke et al. (2007) show that collisions among asteroids were so frequent at those epochs, that there would not be enough *large* bodies left to produce the largest observed impact basins at $t = -3.9$ Gy. Thus, the LHB must have been a late, "cataclysmic" event, which means that we need to find (i) the source and (ii) the dynamical mechanisms responsible for the delay, sudden onset, and short duration of the LHB. From the above discussion it is clear that the solution to this problem must be related to the interaction of the giant planets with the external, massive, planetesimals disc.

In this context, the "smooth" migration model, presented in the previous section, is at least problematic. First, migration ends too soon (~ 100 My from the formation of the planets) to explain the LHB. Second, the final values of eccentricity and mutual inclination of the planets are almost zero. Thus, some excitation mechanism, not present in the "smooth" migration model, must have acted during planetary migration; otherwise the planets would end up on circular and co-planar orbits. These problems are nicely solved by the so-called "Nice model", after the city of Nice (France), in which myself, A. Morbidelli, H. Levison and R. Gomes were all working at the time that the model was developed. The Nice model is presented in the following section.

3.1 The Nice model

There are several unknown parameters that enter a simulation of planet migration. The most important ones are the initial conditions for the planets, and the total mass, surface density profile, and outer edge of the disc. According to Gomes et al. (2004), the mass of the disc must have been between 35 and 50 Earth masses, and its outer edge must have been located at $\sim 30 - 35$ AU. For a more extended and more massive disc, Neptune would undergo runaway migration and would not stop at 30 AU!

When the "Nice" model was being developed, we did not have a way to set constraints on the initial orbital radii of the giant planets – this is no longer true, as we will see in the following. Up to that time, the usual assumption was a well-spread planetary system, with Jupiter starting at ~ 5.5 AU and Neptune at ~ 23 AU. It is true that, if we assume the initial planetary orbits to be as eccentric as today, a more "compact" system would tend to be strongly unstable. However, for initially circular orbits, the system could be stable for billions of years, even if the planets were "packed" within 15 AU from the Sun. We tested this hypothesis by numerical integrations, in which we found that compact systems are indeed stable, provided that the interplanetary distances are not smaller than ~ 3 AU.

An initially compact configuration of the planets evolves in a qualitatively different way, with respect to an initially well-spread one; this is the "key point" in the Nice model. During migration, a pair of planets will have to pass temporarily through a first-order (i.e. the strongest possible) resonance, in which the ratio of their orbital frequencies is nearly constant and equal to a fraction of two small integers, say 1:2, 2:3, 3:5 etc. Then, repeated resonant conjunctions will increase the eccentricity of both orbits. In a well-spread system the planets are far enough from each other to avoid passing through the major resonances. For example, if Jupiter is at 5.5 AU, its 1:2 resonance with Saturn is at 8.73 AU, as given by Kepler's 3rd law. Thus, given that they migrate by moving away from each other, the planets can never cross the 1:2 resonance, if Saturn starts at $a > 8.73$ AU. We did several numerical experiments, changing the initial configuration of the planets, in order to study several resonance crossings. However, we found out that the 1:2 resonance between Jupiter and Saturn would affect the entire system. Thus, at the time, we focused our study on the crossing of the 1:2 Jupiter-Saturn resonance.

A compact planetary system strongly modifies the surface density profile

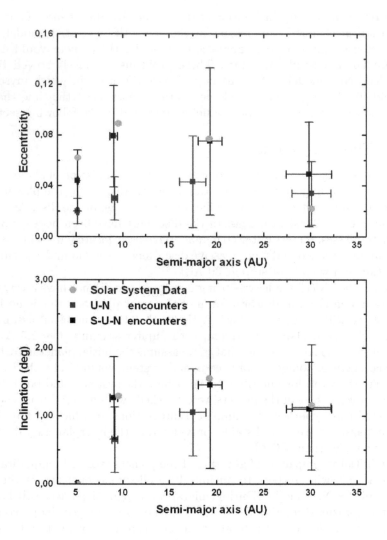

Figure 5. *Statistics of the Nice-model simulations. Squares with error bars correspond to two different classes of runs; only encounters between the ice giants were found in the runs, while encounters with Saturn were also seen in the runs. Both give good statistics, in terms of final a, e and i values, but the runs give a better match to the real solar system data (dots). Reproduced from the results of Tsiganis et al. (2005).*

of the disc. As shown in Figure 2, the dynamical lifetime of disc particles in the interplanetary zone is smaller than the lifetime of the gas disc. Thus, at the beginning of the gas-free era, the compact system of the four giant planets would have already eroded (gravitationally) the inner part of the disc.

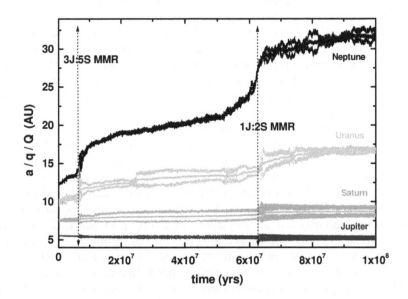

Figure 6. *Time evolution of the semi-major axes, pericentric and apocentric distances of the outer planets in a simulation of planetesimals-driven migration, starting from a multi-resonant configuration. Note the two resonance-crossing episodes. The locations of the resonances are also shown. Reproduced from the results of Morbidelli et al. (2007).*

This implies that planet migration would be driven only by a small flux of particles leaking out from the distant discs, by chaotic motion. Hence, for a realistic planets-disc configuration, as in Figure 2, planetary migration would have been so slow that the planets would not cross the 1:2 resonance before several hundreds of My. For the set of initial conditions shown in Figure 2, the resonance-crossing epoch coincides with the LHB epoch!

After producing these encouraging results, we performed a series of ~ 50 N-body simulations, in which we monitored the evolution of the system, until the depletion of the disc and the end of planet migration (typically ~ 1 Gy). In each simulation the initial positions of the planets, the mass of the disc and the location of the inner edge of the disc were varied. As a result, the observed resonance-crossing time varied between ~ 100 My to 1.2 Gy.

The typical evolution of the planetary system in our runs is shown in Figure 3, where the perihelion (q) and aphelion distances (Q) of each planet are given as functions of time. The difference $Q - q$ is proportional to the eccentricity of the orbit. Figure 4 shows the cumulative mass flux of disc particles that hit the Moon, during the same time interval. As shown in the plot, planet migration is very slow for $\sim$ 880 My, as a result of the small amount of disc particles on planet-crossing orbits. At this point Jupiter and Saturn cross their 1:2 resonance and their orbits become eccentric. The orbits of Uranus and Neptune are then strongly perturbed, due to the small interplanetary distances. In fact their eccentricities grow so much that the orbits of the planets begin to cross, and the planets suffer close encounters with each other. The system is thus led to a short period ($\sim$ 2 My) of chaotic frenzy, during which repeated two- or even three-body encounters increase the eccentricities and mutual inclinations of the planetary orbits.

In 1/3 of our runs, the instability was so strong that one of the ice giants was ejected from the system, by encountering Jupiter. However, in 2/3 of our runs the planetary system was eventually stabilized at the expense of destroying the disc; Uranus and Neptune were forced to penetrate the outer parts of the disc, where dynamical friction reduced their eccentricities and put an end to this planetary billiards. At the same time, disc particles were forced to "fly" all over the solar system. As shown in Figure 4, the amount of comets hitting the Moon increased abruptly at the resonance crossing epoch, and $\sim$ 8.4×10^{21} g of mass accreted on the Moon within $\sim$ 50 My. The above numbers agree well with the time-delay, intensity (estimates give $\sim$ 6×10^{21} g of projectile mass, see Levison et al. 2001) and duration of the LHB. We emphasize that the same series of events occurred in all our successful simulations, since the mechanisms responsible for the onset and suppression of the instability are both deterministic and generic.

At the end of each successful run, the four giant planets were found to follow orbits, very similar to the ones currently observed. Using our set of successful simulations, we computed the mean and standard deviation of the final values of (a,e,i) for all planets. As shown in Figure 5, our simulations fit remarkably well to the real system: the currently observed values of (a,e,i) of *all* giant planets fall within one standard deviation from the mean values found in our runs.

Let us not forget that, in the beginning of the LHB, the inner solar system contained an asteroid belt $\sim$ 10 times more massive than today. We simulated the effect of our chaotic migration model on the asteroid belt, by including in a few simulations 1,000 massless asteroids, with an initial orbital distribution as predicted by Petit et al. (2002). We found that their orbits became unstable, due to *resonance sweeping*: the continuous displacement of resonances throughout the belt, as a result of the migration of Jupiter and Saturn. Our 'asteroid belt' lost $\sim$ 90% of its mass; we seem to have found the missing depletion factor. This process delivers another $3 - 8 \times 10^{21}$ g of asteroid-type "bombs" on the Moon. This result is consistent with recently published results, concerning the composition and size distribution of projectiles that formed the large lunar basins (see Strom et al. 2005).

This chaotic-migration model naturally predicts a cataclysmic bombardment of the Moon, reproducing quite well the main parameters of the LHB. More than that, the "Nice" model explains, for the first time, the orbital

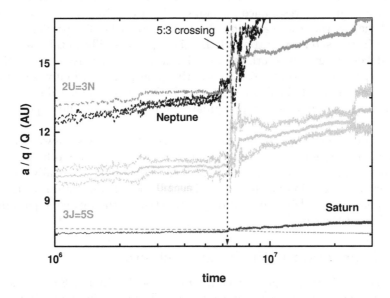

Figure 7. *A close-up of the 5:3 resonance-crossing episode, shown in Figure (6). The resonance lines are shown. Reproduced from the results of Morbidelli et al. (2007).*

architecture of the outer solar system. Although our model is certainly physically plausible, there is no guarantee that these events actually took place. However, there are a number of points that strongly favour our model. To begin with, this model does not involve "exotic" perturbations (e.g. passing stars or rogue planets) of the system. All mechanisms involved are generic to planetary systems. The only critical assumption is that Saturn was formed interior to the location of its 1:2 resonance with Jupiter. So far, no real contradiction between our model and observations has been found. In fact, the Nice model also enables us to explain other interesting issues, as are (i) the absence of collisional asteroid families with ages greater than 3.8 Gy, (ii) the cometary contribution to the Earth's water budget, estimated to $\sim 8 \times 10^{22}$ g ($\sim 6\%$ of the oceans' mass), a value consistent with measurements of the heavy-to-normal (D/H) ratio of ocean water, (iii) the observed mass and orbital distribution of Jupiter Trojans (Morbidelli et al. 2005), and (iv) the trapping of Neptune Trojans.

Recent findings give additional support to the Nice model. Strom et al. (2005) analyzed the size distributions of projectiles responsible for both 'LHB' and 'younger' lunar craters, finding out that the latter ones correspond to the size distribution of the near Earth asteroids, while the former ones to that of main-belt asteroids. Hence, large 'LHB' craters should be formed by asteroids who were ejected from all over the main belt, by a size-independent mechanism, such as resonance sweeping. Moreover, Nesvorný et al. (2007) showed that, during the scattering phase of the planets, disc particles can get trapped in the satellites' region of the giant planets. The total mass of the trapped bodies and their orbital and size distribution agrees well with the observed groups of *irregular satellites* of the giant planets. Finally, very recent results (Levison et al. 2008) suggest that the intricate dynamical structure of the Kuiper Belt could also have originated during the instability epoch of the Nice model.

The Nice model obviously has a long list of successes. However, as we noted in the beginning of the paragraph, a problem remains to be answered: the choice of initial conditions for the planets. The assumed, initially compact, planetary configuration has to agree with the possible final states of the preceding phase of gas-driven migration. As we show in the following paragraph, this is indeed the case.

3.2   Connection with the gas-rich era

As we saw in Section 2, gas-driven migration of the Jupiter-Saturn pair can lead to resonance trapping of the two planets. Then, depending on the parameters of the disc, planet migration can be halted. Morbidelli & Crida (2007) simulated numerically the evolution of Jupiter and Saturn in a MMSN disc and found that, for a wide range of initial conditions, the two planets get trapped in their 2:3 resonance. Thus, for Jupiter stopping at ~ 5.5 AU, Saturn is at ~ 7.3 AU, i.e. much closer to the Sun than suggested in the Nice model. Morbidelli & Crida (2007) showed that, under certain conditions, the planets could be extracted from the resonance, as the gas disc was dissipating. However, in most simulations, Jupiter and Saturn remained locked in resonance even after the dissipation of the disc. This seems to pose a problem for the "Nice" model, but in the opposite sense than one would have expected.

Morbidelli et al. (2007) extended the work of Morbidelli & Crida (2007), by including Uranus and Neptune in the simulations, finding only a limited number of possible outcomes, irrespectively of the exact values of the initial conditions. In particular, Uranus was migrating inwards, getting trapped eventually either in the 3:4 or the 2:3 resonance with Saturn. Similarly, Neptune was migrating inwards, getting trapped either in the 3:4 or 5:6 resonance with Uranus. This result implies that, for a reasonable selection of disc parameters, the four-planets system tends towards one of very few stationary (non-migrating), *multi-resonant* configurations. Note that, the final semi-major axis of Neptune was ~ 13 AU in those simulations; an even more compact configuration than assumed in the Nice model.

Out of the six multi-resonant configurations found, only two remain stable for hundreds of My, thus representing possible "initial conditions" for reproducing the LHB. We performed simulations of planetesimals driven migration,

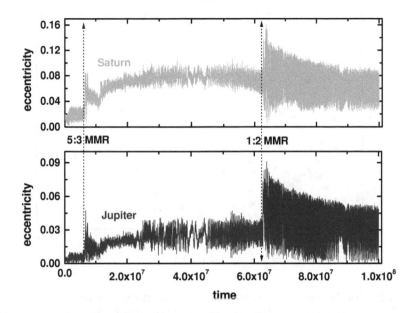

Figure 8. *Eccentricities of Jupiter and Saturn, in the simulation of Figure 6. Resonance crossing leads to the excitation of the planetary orbits. The final values of e are very close to the observed ones for both planets. Reproduced from the results of Morbidelli et al. (2007).*

using these two sets of initial conditions. As shown in Figure 6, the planets begin to migrate, while remaining on nearly circular orbits. At $t \sim 7$ My, the eccentricities of all planets suddenly increase and the migration of (primarily) Neptune and Uranus is accelerated. This migration pattern is very similar to what we observed in the original "Nice"-model simulations of Tsiganis et al. (2005). The difference is that, in this new simulation, the driving mechanism is the crossing of the 3:5 resonance between Jupiter and Saturn (see Figure 7), i.e. the first low-order resonance that Saturn reaches as it migrates outwards, given that its initial position is only ~ 7.6 AU. The 3:5 resonance excites the eccentricities of Jupiter and Saturn (see Figure 8) – which, in turn, excite the eccentricities of Uranus and Neptune – but not as much as the 1:2 resonance. Still, the small – even smaller than in the Nice model – interplanetary distances force the orbits of the ice giant to cross one another, thus leading to repeated encounters and accelerated migration towards the outer parts of the planetesimals' disc. Later on (~ 63 My in the run of Figure 6) Jupiter and Saturn eventually cross the 1:2 resonance, thus acquiring an extra eccentricity "kick" that counterbalances the effects of dynamical friction. The disc is practically dispersed within ~ 100 My, at which point the planets reach their

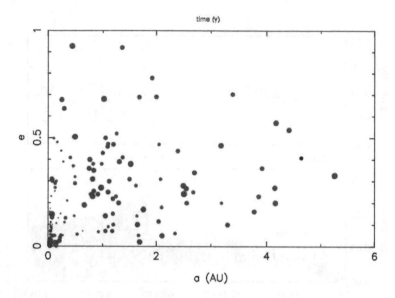

Figure 9. *The typical outcome of the evolution of an unstable multi-resonant state is a single-planet system on an eccentric orbit. Such an "eccentric" Jupiter is shown in this figure (smallest dot-second from right), projected along with a number of observed Extra-solar systems. Reproduced from the results of Morbidelli et al. (2007).*

final orbits. As shown in Figures 6-8, the planetary orbits are quite similar to their currently observed ones.

   This "revised Nice model" is based on exactly the same mechanisms as the original one. The crossing of a resonance increases the eccentricities of the planets and initiates a phase of repeated planetary encounters, which increases mutual inclinations and accelerates the migration. The fact that a different resonance (the 3:5 instead of the 1:2) plays the role of the "trigger" is not important, especially since Jupiter and Saturn also have to cross the 1:2 resonance later on. This second resonance crossing finalizes the orbital eccentricities of the gas giants and is consistent with the findings of Morbidelli et al. (2005) on the capture of Jupiter Trojans. According to our simulations, the duration of the instability phase (roughly, the time interval between the two resonance crossings) is at most ~ 60 My. This fits well to the short duration of the LHB. Finally, we note that the probability of Jupiter having distant encounters with one of the ice giants is larger than in the original "Nice" model, a fact that favors the implantation of disc particles in the irregular satellites' region of Jupiter (Nesvorný et al. 2007).

   To our opinion, these results by Morbidelli et al. (2007) represent a major improvement of the original "Nice" model. The "new" initial conditions for the planets cannot be considered arbitrary any more. They represent, according to this work, the most probable outcome of the gas-driven migration phase of the outer planets. This result supports the idea of an initial compact planetary configuration, that eventually leads to instability through planetesimals-driven migration. We believe that this is a rather complete migration model for the outer solar system, which simultaneously explains a large number of puzzling observations.

4   Implications for extra-solar systems

Migration is a generic process in planetary systems, at least as long as we believe that all planetary systems are formed in a similar way. Thus, the dynamical processes described in the previous sections should have taken place, in a more or less similar way, in all observed extra-solar systems. However, as we saw above, the outcome of each process depends on a multitude of parameters, such as the number of planets, the parameters of the gas disc, the masses of the planets, their order with respect to the host star, etc. Moreover, the stochastic nature of possible instability episodes can lead to a rich variety of final systems. In this final section we show how studying the evolution of the solar system can help us understand the observed – and theoretically expected – diversity of extra-solar planetary systems.

In Section 3 we saw that a 4-planet system (that of the outer planets in the solar system) can reach a stationary, non-migrating, multi-resonant state, under the action of gas-driven migration. However, as we already mentioned, not all multi-resonant configurations are stable. A quadruple resonance is a finely tuned "clock". Small deviations from the exact resonant state can lead to chaotic interactions between the planets. Hence, chaotic scattering among the planets can become possible, leading to a variety of possible outcomes. This phenomenon was indeed observed by Morbidelli et al. (2007) and the final outcome of a representative case is shown here in Figure 9. Scattering among the planets leads to the ejection of all bodies except "Jupiter", which settles on an eccentric orbit. The comparison between this "system" and the observed extra-solar systems places it to the right-hand-side of the diagram. Planet-planet scattering is frequently invoked to explain the large eccentricities of some extra-solar planets. As shown here, the break-up of a multi-resonant state leads naturally to planet-planet scattering and the formation of single-planet (or even multi-planet) eccentric systems, at moderate-to-large orbital distances.

A similar process can occur during planetesimals-driven migration as well. However, more exotic outcomes are also possible, as shown in Figure 10. In this run, all planets survived the resonance-crossing instability, but the two ice giants came very close to each other. Under the action of the planetesimals disc, their orbital eccentricities were suppressed and the planets formed a loose binary in 1:1 resonance, at ~ 48 AU. The relative mean longitude ($\lambda_1 -$

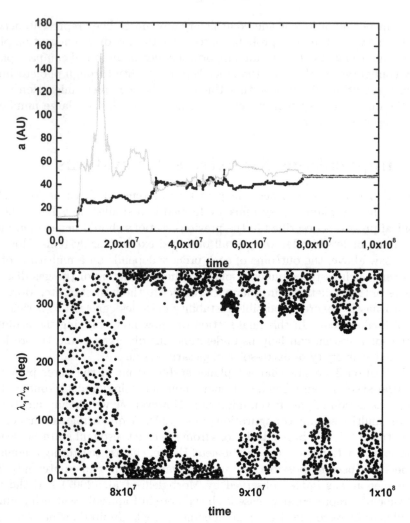

Figure 10. *(top) Evolution of semi-major axis, for the two ice giants that become unstable in this run. Note that, after a series of encounters, both objects have a $\simeq 48$ AU. (bottom) Evolution of the critical argument of the 1:1 resonance, $\lambda_1 - \lambda_2$, during the last 20 My in the simulation. The angle librates around 0, but also, temporarily, around $\pm\pi/3$.*

λ_2) oscillates about zero most of the time, with brief episodes of oscillation about $\pm\pi/3$ being also apparent. Thus, planet-planet scattering inside a disc of planetesimals can lead to the formation of binary (or even Trojan-type) extra-solar planets.

5 Summary

Young planetary systems are inherently "restless". The two phases of planet migration play a decisive role in the final orbital architecture of a system. Given the results presented above, we can attempt to sketch a general path that young planetary systems follow, before reaching maturity.

- During the gas-rich era, gas giants can migrate towards the host star. However, multi-planet systems can follow a different path; they can either migrate towards the star or reach a multi-resonant quasi-stationary state. The latter is the most probable scenario for our solar system.

- As the gas dissipates, multi-resonant systems find themselves in a new environment. The stability of such a configuration is no more guaranteed. Depending on the masses of the planets and the resonance ratios, some configurations can remain stable for very long times, while others can become unstable, leading to planet-planet scattering and highly eccentric planets.

- Stable, multi-resonant, systems can undergo planetesimals-driven migration; this process can be very slow, depending on the parameters of the disc. This is a quite favorable scenario for the solar system, as it is the only way to associate the subsequent instability phase with the LHB. During this slow migration phase, terrestrial planets could form in the inner parts of the system.

- After a possibly large amount of time two planets may be forced to cross a mean motion resonance. This can provoke a global instability of the system, which could be suppressed at the expense of destroying the disc, or even losing a planet.

- Such an instability phase, would be the final episode in the early life of a planetary system. The orbital configuration of the planets and small bodies' populations would be completely reshaped. After that, the system reaches maturity – the planetary orbits no longer change.

Although we believe the general lines of the above scheme to be roughly correct, it is certainly not the complete picture. A lot of work is needed to further understand in detail the exact mechanisms of planet-disc interactions. New observations and theoretical work on extra-solar systems may lead us to revise some of the above ideas in the near future. It is certainly a fascinating time to work on planetary systems!

Acknowledgments

I wish to thank the organizers of the Skye school for inviting me to participate in this very interesting school, for their hospitality and for giving me a chance to visit such an amazing place.

References

Armitage, P J, 2007, *Lecture notes on the formation and early evolution of planetary systems*, ArXiv Astrophysics e-prints arXiv:astro-ph/0701485

Boss A P, 1997, Giant planet formation by gravitational instability. *Science* **276** 1836

Boss A P, 2000, Possible rapid gas giant planet formation in the solar nebula and other protoplanetary disks. *Astrophysical Journal* **536** L101

Bottke W F, Levison H F, Nesvorný D and Dones L, 2007, Can planetesimals left over from terrestrial planet formation produce the lunar Late Heavy Bombardment? *Icarus* **190** 203

Cohen B A, Swindle T D, Kring D A, 2000, Support for the lunar cataclysm hypothesis from lunar meteorite impact melt ages, *Science* **290** 1754

Crida A, Morbidelli A and Masset F, 2006, On the width and shape of gaps in protoplanetary disks, *Icarus* **181** 587

Dalrymple G B and Ryder G, 1993, Ar-40/Ar-39 age spectra of Apollo 15 impact melt rocks by laser step-heating and their bearing on the history of lunar basin formation, *Journal of Geophysical Research* **98** 13085

Fernandez J A and Ip W-H, 1984, Some dynamical aspects of the accretion of Uranus and Neptune - The exchange of orbital angular momentum with planetesimals, *Icarus* **58** 109

Goldreich P and Tremaine S, 1979, The excitation of density waves at the Lindblad and corotation resonances by an external potential, *Astrophysical Journal* **233** 857

Goldreich P and Tremaine S, 1980, Disk-satellite interactions, *Astrophysical Journal* **241** 425

Gomes R S, Morbidelli A and Levison H F, 2004, Planetary migration in a planetesimal disk: why did Neptune stop at 30 AU?, *Icarus* **170** 492

Gomes R, Levison H F, Tsiganis K and Morbidelli A, 2005, Origin of the cataclysmic Late Heavy Bombardment period of the terrestrial planets, *Nature* **435** 466

Ida S, Bryden G, Lin D N C and Tanaka H, 2000, Orbital migration of Neptune and orbital distribution of Trans-Neptunian objects, *Astrophysical Journal* **534** 428

Jewitt D and Luu J, 1993, Discovery of the candidate Kuiper belt object 1992 QB1, *Nature* **362** 730

Kokubo E and Ida S, 2000, Formation of protoplanets from planetesimals in the solar nebula, *Icarus* **143** 15

Levison H F, Dones L, Chapman C R, Stern S A, Duncan M J and Zahnle K, 2001, Could the lunar "Late Heavy Bombardment" have been triggered by the formation of Uranus and Neptune?, *Icarus* **151** 286

Levison H F, Morbidelli A, Gomes R and Backman D, 2007, Planet migration in planetesimal Disks, in *Protostars and Planets V*, editor Reipurth B, Jewitt D and Keil K (University of Arizona Press, Tucson)

Levison H F, Morbidelli A, Vanlaerhoven C, Gomes R and Tsiganis K, 2008, Origin of the structure of the Kuiper belt during a dynamical instability in the orbits of Uranus and Neptune, *Icarus* **196** 258

Malhotra R, 1995, The origin of Pluto's orbit: Implications for the solar system beyond Neptune, *Astronomical Journal* **110** 420

Masset F S and Papaloizou J C B, 2003, Runaway migration and the formation of Hot Jupiters, *Astrophysical Journal* **588** 494

Masset F S, Morbidelli A, Crida A and Ferreira J, 2006, Disk surface density transitions as protoplanet traps, *Astrophysical Journal* **642** 478

Morbidelli A, Levison H F, Tsiganis K and Gomes R, 2005, Chaotic capture of Jupiter's Trojan asteroids in the early solar system, *Nature* **435** 462

Morbidelli A and Crida A, 2007, The dynamics of Jupiter and Saturn in the gaseous protoplanetary disk, *Icarus* **191** 158

Morbidelli A, Tsiganis K, Crida A, Levison H F and Gomes R, 2007, Dynamics of the giant planets of the solar system in the gaseous protoplanetary disk and their relationship to the current orbital architecture, *Astronomical Journal* **134** 1790

Morbidelli A, Crida A, Masset F and Nelson R P, 2008, Building giant-planet cores at a planet trap, *Astronomy and Astrophysics* **478** 929

Nesvorný D, Vokrouhlický D and Morbidelli A, 2007, Capture of irregular satellites during planetary encounters, *Astronomical Journal* **133** 1962

Papaloizou J C B, Nelson R P, Kley W, Masset F S, Artymowicz P, 2007, Disk-planet interactions during planet formation, in *Protostars and Planets V*, editor Reipurth B, Jewitt D and Keil K (University of Arizona Press, Tucson)

Papaloizou J C B and Terquem C, 2006, Planet formation and migration. *Reports of Progress in Physics* **69** 119

Petit J-M, Chambers J, Franklin F and Nagasawa M, 2002, Primordial excitation and depletion of the main belt, in *Asteroids III*, editor Bottke W F, Paolicchi P, Binael R P and Cellino A (University of Arizona Press, Tucson)

Pierens A and Nelson R P, 2008, On the formation and migration of giant planets in circumbinary discs, *Astronomy and Astrophysics* **483** 633

Pollack J B, Hubickyj O, Bodenheimer P, Lissauer J J, Podolak M and Greenzweig Y, 1996, Formation of the giant planets by concurrent accretion of solids and gas, *Icarus* **124** 62

Shakura N I and Syunyaev R A, 1973, Black holes in binary systems. Observational appearance, *Astronomy and Astrophysics* **24** 337

Strom R G, Malhotra R, Ito T, Yoshida F and Kring D A, 2005, The origin of planetary impactors in the inner solar system, *Science* **309** 1847

Syer D and Clarke C J, 1995, Satellites in discs: regulating the accretion luminosity, *Monthly Notices of the Royal Astronomical Society* **277** 758

Tanaka H, Takeuchi T and Ward W R, 2002, Three-Dimensional interaction between a planet and an isothermal gaseous disk. I. Corotation and Lind-blad Torques and planet migration, *Astrophysical Journal* **565** 1257

Tera F, Papanastassiou D A and Wasserburg G J, 1974, Isotopic evidence for a terminal lunar cataclysm, *Earth and Planetary Science Letters* **22** 1

Thommes E W and Lissauer J J, 2003, Resonant inclination excitation of migrating giant planets, *Astrophysical Journal* **597** 566

Tsiganis K, Gomes R, Morbidelli A, Levison H F, 2005, Origin of the orbital architecture of the giant planets of the Solar System, *Nature* **435** 459

Ward W R, 1997, Protoplanet migration by Nebula tides, *Icarus* **126** 261

Part III

Dynamics of Planetary Systems

A brief account of mutual planetary perturbations

PJ Message

Applied Mathematics Division, Mathematical Sciences Department,
University of Liverpool L69 3BX, U.K.

This chapter gives a very brief review of some methods for describing motions in a planetary system, including the perturbations which arise from their mutual gravitational attractions, and for deriving expressions of these in the form of asymptotic multiple Fourier series. It is shown how, with an appropriate choice of parameters to describe the orbits, these expressions contain only periodic terms, some however of very long periods (in the solar system, some of many millions of years).

1   Introduction

Expressions derived by the methods described here, or by equivalent methods, have had undoubted success in predicting planetary positions in the solar system, as verified by observations, over many centuries, and by comparison with numerical integrations of the equations of motion, over at least many hundreds of thousands of years. This success is all the more remarkable in the face of the obstacles which arise, from the everywhere-dense commensurabilities of mean orbital motions, to any uniform convergence of the asymptotic series used, and in turn, to validation of the series by the usual methods of rigorous mathematical analysis. We will see how the effect of these commensurabilities is tempered by the "d'Alembert Property" of the expansions, because the orbits of the planets are very nearly coplanar, and of small eccentricity, though this fails where there is a very close small-integer commensurability, when the solution takes a different form. We may expect the same success in many other planetary systems, especially where the orbits are also nearly coplanar and nearly circular.

2 Review of Kepler two-body motion

2.1 The orbit in general

In a planetary or satellite system, in which most of the mass is concentrated in one body (the Sun, in the solar system, or the primary planet, in a satellite system), we use the concept of the *instantaneous Kepler orbit* (usually an ellipse) of each planet (or satellite) to study the evolution of the system, since the parameters which define these orbits are usually subject to slow changes (*perturbations*), in response to the attractions of the other members of the system. So, to set up a set of such parameters, we begin by reviewing the simpler case of a system with just one planet, P, of mass m, moving around the primary, S, of mass M. If $\mathbf{r}$ is the relative position vector from S to P, and r its length, then the equation governing their relative motion, taking into account their mutual gravitational attractions, is

$$\frac{\mathrm{d}^2\mathbf{r}}{\mathrm{d}t^2} = -\frac{GM}{r^2}(\mathbf{r}/r) - \frac{Gm}{r^2}(\mathbf{r}/r)$$

(where G is the constant of gravitation) or, more concisely,

$$\frac{\mathrm{d}^2\mathbf{r}}{\mathrm{d}t^2} = -\mu(\mathbf{r}/r^3), \tag{1}$$

where μ is $G(M+m)$. From this we see that

$$\frac{\mathrm{d}(\mathbf{r} \times \dot{\mathbf{r}})}{\mathrm{d}t} = \mathbf{0},$$

so that $\mathbf{r} \times \dot{\mathbf{r}}$ is a constant vector, $\mathbf{h}$, the "angular momentum per unit mass" in the relative motion. We see that $\mathbf{r}$ is always perpendicular to $\mathbf{h}$, so that the motion is planar. Now we see that

$$
\begin{aligned}
\mathbf{h} \times \frac{\mathrm{d}^2\mathbf{r}}{\mathrm{d}t^2} &= \mathbf{h} \times (-\mu\mathbf{r}/r^3) \\
&= -\frac{\mu}{r^3}((\mathbf{r} \times \dot{\mathbf{r}}) \times \mathbf{r}) \\
&= -\frac{\mu}{r^3}(\dot{\mathbf{r}}r^2 - \mathbf{r}r\dot{r}) \\
&= -\frac{\mu\dot{\mathbf{r}}}{r} + \frac{\mu\mathbf{r}\dot{r}}{r^2} \\
&= -\frac{\mathrm{d}}{\mathrm{d}t}(\mu\mathbf{r}/r)
\end{aligned}
$$

so that

$$\mathbf{h} \times \dot{\mathbf{r}} = -\mu(\mathbf{r}/r + \mathbf{e}), \tag{2}$$

where **e** is a constant vector, perpendicular to **h**, and so in the plane of the motion. Also

$$\mathbf{r} \cdot (\mathbf{h} \times \dot{\mathbf{r}}) \;=\; -\mathbf{h} \cdot (\mathbf{r} \times \dot{\mathbf{r}})$$
$$=\; -h^2$$

(where h is the modulus of **h**), so that

$$h^2 \;=\; \mu\mathbf{r} \cdot (\mathbf{r}/r + \mathbf{e})$$
$$=\; \mu(r + \mathbf{e} \cdot \mathbf{r}) \tag{3}$$

and thus

$$r(1 + e\cos f) \;=\; h^2/\mu$$
$$=\; p \tag{4}$$

say, where f, called the *true anomaly*, is the angle between **r** and **e**, and e is the modulus of **e**. This is the equation of a conic section, of eccentricity e, and with semilatus rectum $p = h^2/\mu$. The minimum distance of P from S (sometimes denoted by q), where $f = 0$, is thus $p/(1 + e)$, at the point P_c, which is the *pericentre*, or *perihelion* (if the primary is the Sun), or *perigee* (if the primary is the Earth). Thus the vector **e** is towards the pericentre.

2.2   The elliptic orbit

In the case $0 \le e < 1$, i.e. the *ellipse* (which includes the circle, for $e = 0$), there is a maximum distance (sometimes denoted by q'), where $f = \pi$, which is thus $p/(1 - e)$, at the point P_a, which is the *apocentre*, or *aphelion*, or *apogee*. The line joining the pericentre (P_c) to the apocentre (P_a) is called the *major axis*, of length

$$p/(1 + e) + p/(1 - e) \;=\; \frac{2pe}{1 - e^2}$$
$$=\; 2a$$

so that half its length is a, the *major semi-axis*, and then

$$p \;=\; a(1 - e^2) \tag{5}$$

and so $q = a(1 - e)$, and $q' = a(1 + e)$. The midpoint, C, of the major axis, is the *centre* of the ellipse, and its distance from the *focus*, S, is $CS = ae$.

Consider the *auxiliary circle*, which is the circle on the major axis, $P_c\,P_a$, as diameter, and take the point Q on it such that QP is perpendicular to the major axis. Then the angle $P_c\,C\,Q$ is called the *eccentric anomaly* (E) . Let N be where QP meets the major axis. Then equating the expression for SN in terms of r and f, as the projection of SP onto the major axis, to that in terms of a and E by projecting CQ onto the major axis and subtracing CS, we get $r \cos f = a(\cos E - e)$. Then, from the Eq. (4) of the ellipse,

$$
\begin{aligned}
r &= p - er \cos f \\
&= a\,(1 - e \cos E)
\end{aligned}
\tag{6}
$$

Using these last two results, we deduce that $r \sin f = b \sin E$, where $b^2 = a^2(1 - e^2)$. Then b, the *minor semi-axis*, is half of the *minor axis*, which, passing through the centre, C, and perpendicular to the major axis, joins opposite sides of the ellipse. We note that the distances of Q and P from the major axis are in the ratio $a : b$, so that the ellipse may be regarded as the circle compressed in the ratio $b : a$ in the direction perpendicular to the major axis.

2.3   The orbit in time

Now to consider the position in the orbit at a given time, note that the transverse component of $\mathbf{r} \times \dot{\mathbf{r}} = \mathbf{h}$, gives $r^2 \dot{f} = h$, showing that the line SP covers area at a constant rate. (This is Kepler's second law of planetary motion.) Differentiating the equation of the ellipse Eq. (4), and using the last result, leads to

$$
\dot{r} = he \sin f / p
\tag{7}
$$

Differentiating the Eq. (6) gives

$$
\dot{r} = ae \sin E\, \dot{E}
$$

and equating the two expressions for $\dot{r}$ gives $r\dot{E} = hb/(ap)$, whence

$$
(1 - e \cos E)\dot{E} = n
$$

where $n = hb/(a^2 p) = h/(ab)$, so we get *Kepler's equation*:

$$
(E - e \sin E) = \ell
\tag{8}
$$

where ℓ, the *mean anomaly*, is an angle which increases uniformly at the rate n (the *mean motion*), in undisturbed Kepler two-body motion, and takes the value zero (modulo 2π) at pericentre, and the value π (modulo 2π) at apocentre. Note that $n^2 a^3 = \mu$, which is Kepler's third law of planetary motion. From Kepler's equation we derive the Fourier series for E:

$$
E = \ell + 2 \sum_{s=1}^{\infty} (J_s(se)/s) \sin s\ell
\tag{9}
$$

where $J_n(x)$ is *Bessel's function*, given by $\sum_{r=0}^{\infty} \frac{(-1)^r (x/2)^{n+2r}}{r!(n+r)!}$. We also derive other expansions, including

$$
r = a\{1 + e^2/2 - 2e \sum_{s=1}^{\infty} J_s'(se)/s \cos s\ell\}
$$

$$= a\{1 + e^2/2 - (e - 3e^3/8 + 5e^5/192 \ldots)\cos\ell$$
$$- (e^2/2 - e^4/3 \ldots)\cos 2\ell$$
$$- (3e^3/8 - 45e^5/128 \ldots)\cos 3\ell$$
$$- (e^4/3 \ldots)\cos 4\ell$$
$$- (125e^5/384 \ldots)\cos 5\ell \ldots\}$$

and

$$f = \ell + (2e - e^3/4 \ldots)\sin\ell$$
$$+ (5e^2/4 \ldots)\sin 2\ell$$
$$+ (13e^3/12 \ldots)\sin 3\ell \ldots$$

The latter is the *equation of the centre*. In each of these expansions, note the *d'Alembert property*, that the coefficient of each term is a power series in alternating powers of e, the lowest power being equal to the multiple of ℓ in the term. This is equivalent to the expressibility of each expansion in positive powers of $e\cos\ell$ and $e\sin\ell$.

2.4 The orientation of the orbit in space

The *inclination* (i) of the orbit is the angle between the plane of the orbit and the reference plane. In the solar system, the reference plane is usually the *ecliptic plane*, which is the plane of the Sun's apparent annual motion about the Earth, which should be specified as at a given date, because of the movement of that plane in response to planetary perturbations. The *ascending node* is the line of intersection of the orbit plane with the reference plane, usually in the sense towards the point at which the planet passes from the southern to the northern side. The *longitude of the ascending node* (Ω) is the angle, measured in the reference plane in the agreed positive sense (usually that of the planetary motions), to the ascending node from the reference direction. In the solar system, the reference direction is usually the *First Point of Aries*, which is the intersection of the ecliptic plane with the plane of the Earth's equator, in the direction of the Sun when it appears from Earth to cross the equator from south to north, in its annual motion about the Earth, at the *spring equinox*. (This direction should also be specified as at a given date). The *argument of pericentre* (or *of perihelion*, or *of perigee*.) (ω) is the angle, in the orbit plane, from the ascending node to the direction of the pericentre. When the inclination is small, the line of the node, and consequently Ω and ω, may be badly determined, and so we often use the *longitude of pericentre*, $\varpi = \Omega + \omega$ (sometimes called a "broken angle", because it is the sum of two angles measured in different planes). It will be better determined and remains defined even when the inclination is zero! When the eccentricity is small, the direction of the major axis may also be badly determined, and so also may be ω and ℓ, and so we often use the *mean longitude* in the orbit ($\lambda = \varpi + \ell$), which remains defined even when the eccentricity is zero.

3   Perturbed elliptic motion

In systems with more than one planet, but where the primary body has by far the greatest mass, it is often convenient to use the Keplerian elliptic motions of the component planets, relative to the primary, as a first approximation to the motion, and to regard the mutual attractions of the planets as causing changes (*perturbations*) in the Kepler ellipses. We are effectively using, as parameters to describe the position and velocity of each planet relative to the primary at a given instant, not components of the position relative vector $\mathbf{r}$ and velocity $\dot{\mathbf{r}}$, but the parameters $(a, e, i, \lambda, \varpi, \Omega)$ of that Kepler orbit, called the *instantaneus orbit*, which corresponds to the position and velocity at that instant. These parameters (apart from the mean longitude λ) will usually be changing much more slowly than the co-ordinates and components of velocity, and the mean longitude will usually be increasing nearly linearly in time, as it would be in undisturbed Kepler motion). Let us set up the equations governing perturbations of planetary orbits arising from mutual attractions. Suppose we have a system of n planets, P_1, P_2, ..., P_n, with primary S, of mass M, the mass of P_i being m_i. Let the position vector, in an inertial frame, of S be ρ_S, and of P_i be ρ_i. Then the equations of motion are, in the inertial frame,

$$\frac{\mathrm{d}^2\rho_i}{\mathrm{d}t^2} = -\frac{GM}{r_i^2}(\mathbf{r_i}/r_i) - \sum_{j=1 \ \neq i}^{n} \left(\frac{Gm_j}{r_{ij}^2}(\mathbf{r_i} - \mathbf{r_j})/r_{ij} \right)$$

and

$$\frac{\mathrm{d}^2\rho_S}{\mathrm{d}t^2} = -\sum_{j=1}^{n} \frac{Gm_j}{r_j^2}(\mathbf{r_j}/r_j)$$

where $\mathbf{r_i} = \rho_i - \rho_S$ is the position of P_i relative to S, r_i is the modulus of $\mathbf{r_i}$, and r_{ij} that of $\mathbf{r_i} - \mathbf{r_j}$. From these,

$$\frac{\mathrm{d}^2\mathbf{r_i}}{\mathrm{d}t^2} = -\frac{\mu_i \mathbf{r_i}}{r_i^3} - \sum_{j=1 \ \neq i}^{n} Gm_j \left(\frac{(\mathbf{r_j} - \mathbf{r_i})}{r_{ij}^3} - \frac{\mathbf{r_j}}{r_j^3} \right)$$

where here μ_i is $G(M + m_i)$. This may be written

$$\frac{\mathrm{d}^2\mathbf{r_i}}{\mathrm{d}t^2} = -\frac{\mu_i \mathbf{r_i}}{r_i^3} + \frac{\partial R_i}{\partial \mathbf{r_i}} \tag{10}$$

where

$$R_i = \sum_{j=1 \ \neq i}^{n} Gm_j \left(\frac{1}{r_{ij}} - \frac{\mathbf{r_i} \cdot \mathbf{r_j}}{r_j^3} \right) \tag{11}$$

which is called the *disturbing function*. The changes in the orbital parameters of a planet are governed by *Lagrange's planetary equations*, which for a planet with orbital parameters $(a, e, i, \lambda, \varpi, \Omega)$ subject to perturbations derivable from a disturbing function R, are:

$$
\begin{aligned}
\frac{da}{dt} &= \frac{2}{na}\frac{\partial R}{\partial \lambda} \\[4pt]
\frac{de}{dt} &= -\frac{b}{na^3 e}\frac{\partial R}{\partial \varpi} - \frac{b(a-b)}{na^4 e}\frac{\partial R}{\partial \lambda} \\[4pt]
\frac{di}{dt} &= -\frac{1}{nab\sin i}\frac{\partial R}{\partial \Omega} - \frac{\tan(i/2)}{nab}\left(\frac{\partial R}{\partial \varpi} + \frac{\partial R}{\partial \lambda}\right) \\[4pt]
\frac{d\lambda}{dt} &= n - \frac{2}{na}\frac{\partial R}{\partial a} + \frac{b(a-b)}{na^4 e}\frac{\partial R}{\partial e} + \frac{\tan(i/2)}{nab}\frac{\partial R}{\partial i} \\[4pt]
\frac{d\varpi}{dt} &= \frac{b}{na^3 e}\frac{\partial R}{\partial e} + \frac{\tan(i/2)}{nab}\frac{\partial R}{\partial i} \\[4pt]
\frac{d\Omega}{dt} &= \frac{1}{nab\sin i}\frac{\partial R}{\partial i}
\end{aligned}
\tag{12}
$$

The structure of these equations shows that we are quite close to a canonical system, as we shall see later. For more general types of perturbation, if the component of the acceleration in the outward radial direction is $-\frac{\mu r}{r^3} + S$, the component in the transverse direction (in the orbit plane) is T, and the component perpendicular to the orbit plane is W, then the changes in the orbital parameters of the planet are governed by *Gauss' planetary equations*:

$$
\begin{aligned}
\frac{da}{dt} &= \frac{2}{n}\left(\frac{ae\sin f}{b}S + \frac{b}{r}T\right) \\[4pt]
\frac{de}{dt} &= -\frac{b}{na^2}\left(\sin f\, S + \frac{b}{ae}T\right) \\[4pt]
\frac{di}{dt} &= -\frac{r\cos u}{nab}W \\[4pt]
\frac{d\Omega}{dt} &= \frac{r\sin u}{nab\sin i}W \\[4pt]
\frac{d\varpi}{dt} &= \frac{1}{na}\left(\frac{-b}{ae}\cos u\, S + \frac{r\sin f}{be}(2 + e\cos f)T + \frac{r\sin u\tan(i/2)}{b}W\right) \\[4pt]
\frac{d\lambda}{dt} &= n + \frac{1}{na}\left(\frac{-b}{ae}\cos u\, S + \frac{(a-b)r\sin f(2 + e\cos f)}{abe}T\right) \\[4pt]
&\quad + \frac{1}{na}\left(\frac{r\sin u\tan(i/2)}{b}W\right)
\end{aligned}
\tag{13}
$$

Here $u = \omega + f$ is the *argument of latitude*, the angle from the line of the ascending node to SP_i.

3.1   The main features of the disturbing function

To use Lagrange's equations, we need the disturbing function R expressed in terms of the parameters $(a, e, i, \lambda, \varpi, \Omega)$ of the planet P. Suppose that R' is the part of R arising from the action of another planet, P', of mass m', and with orbital parameters $(a', e', i', \lambda', \varpi', \Omega')$. To derive some important properties of R', let us first note that it is of period 2π in each of $(\lambda, \varpi, \Omega, \lambda', \varpi', \Omega')$, and so may be expressed as a Fourier series in each of these, so we may write

$$R' = \sum_j (K_j \cos N_j + L_j \sin N_j)$$

where $N_j = j_1\lambda + j_2\varpi + j_3\Omega + j_4\lambda' + j_5\varpi' + j_6\Omega'$, and the summation is over all sets of integers $j = (j_1, j_2, j_3, j_4, j_5, j_6)$, where we may suppose without loss of generality that $j_1 \geq 0$. Consider the transformation given by

$$\lambda \mapsto -\lambda, \varpi \mapsto -\varpi, \Omega \mapsto -\Omega, \lambda' \mapsto -\lambda', \varpi' \mapsto -\varpi', \Omega' \mapsto -\Omega'$$

This leaves the relative positions of P and P' unchanged, and so leaves the distances r, r', and r_{ij} and the scalar product $\mathbf{r} \cdot \mathbf{r}'$ unchanged, and so leaves R' unchanged, but replaces N_j by $-N_j$, and so

$$\sum_j (K_j \cos N_j + L_j \sin N_j) = \sum_j (K_j \cos N_j - L_j \sin N_j)$$

so that $\sum_j (L_j \sin N_j) = 0$. But this is an identity in the $(\lambda, \varpi, \Omega, \lambda', \varpi', \Omega')$, and so $L_j = 0$ for each j. Thus $R' = \sum_j K_j \cos N_j$. Further, suppose the configuration is rotated through an angle χ about the perpendicular to the reference plane, so that

$$\lambda \mapsto \lambda + \chi, \varpi \mapsto \varpi + \chi, \Omega \mapsto \Omega + \chi, \lambda' \mapsto \lambda' + \chi, \varpi' \mapsto \varpi' + \chi, \Omega' \mapsto \Omega' + \chi$$

Once again, relative positions, and so R', are unchanged, so that

$$\sum_j K_j \cos N_j = \sum_j K_j \cos \left(N_j + (\sum_{i=1}^6 j_i)\chi\right)$$

But this is an identity in the $(\lambda, \varpi, \Omega, \lambda', \varpi', \Omega')$, and so $\sum_j K_j \sin \left((\sum_{i=1}^6 j_i)\chi\right) = 0$, which in turn is an identity in χ, so that $K_j = 0$ unless $\sum_{i=1}^6 j_i = 0$. Further, the d'Alembert property is preserved through the algebraic processes in the construction of R', so we find that R' may be expressed in positive powers of $e \cos \varpi$, $e \sin \varpi$, $e' \cos \varpi'$, $e' \sin \varpi'$, and also of $\sin i \cos \Omega$, $\sin i \sin \Omega$, $\sin i' \cos \Omega'$, and $\sin i' \sin \Omega'$. Therefore K_j has the factor $e^{|j_2|} \sin^{|j_5|} i \, e'^{|j_3|} \sin^{|j_6|} i'$.

4 The canonical form of the equations of motion

There now follows a summary of a method for the solution of the equations for the mutual perturbations in a planetary system, which will use Hamiltonian canonical form. In this, the state of motion of the system at any time is specified by a set of *generalised co-ordinates*,

$$q = (q_1, q_2, \ldots, q_n)$$

and also a set of *generalised momenta*,

$$p = (p_1, p_2, \ldots, p_n)$$

and the equations of motion may be written as *Hamilton's equations*,

$$\frac{\mathrm{d}q_i}{\mathrm{d}t} = \frac{\partial H}{\partial p_i} \quad \text{and} \quad \frac{\mathrm{d}p_i}{\mathrm{d}t} = -\frac{\partial H}{\partial q_i}$$
$$\text{for} \quad i = 1, 2, \ldots, n. \tag{14}$$

where H is the *Hamiltonian function*, and n is the number of *degrees of freedom* of the system. It follows that, for any function $f(q, p)$,

$$\frac{\mathrm{d}f}{\mathrm{d}t} = \{f, H\} \tag{15}$$

where the *Poisson bracket* of two functions f and g is given by

$$\{f, g\} = \sum_{i=1}^{n} \left(\frac{\partial f}{\partial q_i} \frac{\partial g}{\partial p_i} - \frac{\partial f}{\partial p_i} \frac{\partial g}{\partial q_i} \right)$$

4.1 Lie series transformations

If $W(q, p)$ is any function with the necessary derivatives, define $\mathcal{L}_W$ to be the operator taking any function $f(q, p)$ to its Poisson bracket with W, that is

$$\mathcal{L}_W(f) = \{f, W\} \tag{16}$$

and we write repeated applications of it as $\mathcal{L}_W^{k+1}(f) = \mathcal{L}\left(\mathcal{L}_W^k(f)\right)$, understanding $\mathcal{L}_W^1(f)$ as $\mathcal{L}_W(f)$. Then, if ϵ is any real number, consider the transformation

$$(q, p) \mapsto (Q, P) \tag{17}$$

(where Q is $(Q_1, Q_2, \ldots, Q_n)$, and P is $(P_1, P_2, \ldots, P_n)$) defined by

$$Q_i = q_i + \sum_{j=1}^{\infty} \frac{\epsilon^j}{j!} \mathcal{L}_W^j(q_i) \tag{18}$$

and

$$P_i = p_i - \sum_{j=1}^{\infty} \frac{\epsilon^j}{j!} \mathcal{L}_W^j(p_i) \tag{19}$$

so, for any $f(q,p)$,

$$f(Q,P) = f(q,p) \sum_{j=1}^{\infty} \frac{\epsilon^j}{j!} \mathcal{L}_W^j\left(f(q,p)\right). \tag{20}$$

This transformation may be regarded as representing the evolution over a time interval of length ϵ of a hypothetical dynamical system in which the Hamiltonian function is W. It follows that Eq. (17) is a contact transformation. Thus (Q,P), like (q,p), are governed, in the actual motion, by equations of motion of Hamilton's type,

$$\begin{aligned} \dot{Q}_i &= \frac{\partial H'}{\partial P_i}, \\ \text{and} \quad \dot{P}_i &= -\frac{\partial H'}{\partial Q_i}, \\ \text{for} \quad i &= 1, 2, ..., n. \end{aligned} \tag{21}$$

for an appropriate Hamiltonian function $H'(Q,P)$. The transformation (17), so defined, is a *Lie series transformation*, and the function W is its *generating function*, the choice of which we may make according to what we want the transformation to do. Note that, having regard to (15), for any function f,

$$f(Q,P) = f(q,p) + \sum_{k=1}^{\infty} \frac{1}{k!} \epsilon^k \mathcal{L}_W^k\left(f(q,p)\right) \tag{22}$$

The inverse of the transformation is given by reversing the sign of ϵ, that is

$$\begin{aligned} q_i &= Q_i + \sum_{k=1}^{\infty} \frac{1}{k!} (-\epsilon)^k \mathcal{L}_W^k(Q_i), \\ \text{and} \quad p_i &= P_i + \sum_{k=1}^{\infty} \frac{1}{k!} (-\epsilon)^k \mathcal{L}_W^k(P_i), \\ \text{for} \quad i &= 1, 2, ..., n, \end{aligned} \tag{23}$$

and

$$f(q,p) = f(Q,P) + \sum_{k=1}^{\infty} \frac{1}{k!} (-\epsilon)^k \mathcal{L}_W^k\left(f(Q,P)\right) \tag{24}$$

4.2 A canonical formulation of an n-planet system

Suppose, as before, that we have a system of n planets, P_1, P_2, ..., P_n, with primary S, of mass M, the mass of P_i being m_i. We will use the *Jacobi* system of relative position vectors, which makes it easier to set up the equations in canonical form, and so reap the advantages of the use of contact transforms, as we will see. In this system, the position vector of P_1 relative to S is $\mathbf{r}_1$, that of P_2 relative to $G_{S,1}$ (the mass-centre of S and P_1) is $\mathbf{r}_2$, and so on, that of P_i relative to $G_{S1,2,...,i-1}$ (the mass-centre of S, P_1, ..., and P_{i-1}) being $\mathbf{r}_i$. Let ϵ be the greatest of the m_i/M, and let the β_i be defined by $\epsilon\beta_i = m_i M_{i-1}/M_i$, where M_i is $M + \sum_{k=1}^{i} m_k$. Then $m_i = \frac{\epsilon\beta_i}{(1-\epsilon\beta_i/M_{i-1})}$ and we find that the equations of motion take the form

$$\beta_i \frac{d^2\mathbf{r}_i}{dt^2} = -\mu_i\beta_i\mathbf{r}_i/r_i^3 + \frac{\partial R}{\partial \mathbf{r}_i}$$

for $i = 1, 2, \ldots, n$, where r_i is the modulus of $\mathbf{r}_i$, μ_i is defined by $\epsilon\mu_i\beta_i = GMm_i$, and in fact $\mu_i = G\{M + \epsilon\beta_i + O(\epsilon^2)\}$. Also R is a function of the relative positions and masses which tends to zero with ϵ, and so serves as disturbing function. Let $\mathbf{p}_i = \beta_i\dot{\mathbf{r}}_i$, for $i = 1, 2, \ldots, n$, then the $\mathbf{r} = (\mathbf{r}_1, \mathbf{r}_2, \ldots, \mathbf{r}_n)$ are canonical co-ordinates with the $\mathbf{p} = (\mathbf{p}_1, \mathbf{p}_2, \ldots, \mathbf{p}_n)$ as their conjugate momenta, respectively, the Hamiltonian function being

$$H(\mathbf{r}, \mathbf{p}) = \sum_{i=1}^{n} \left(\frac{\mathbf{p}_i^2}{2\beta_i} - \frac{\mu_i\beta_i}{r_i} \right) - R$$

Let now $(a_i, e_i, i_i, \lambda_i, \varpi_i, \Omega_i)$ be the orbital elements of the instantaneous orbit corresponding to $\mathbf{r}_i$ and $\dot{\mathbf{r}}_i = \mathbf{p}_i/\beta_i$. Then put $q_i^\dagger = \lambda_i$, $q_{n+i}^\dagger = \varpi_i$, and $q_{2n+i}^\dagger = \Omega_i$, and take these as canonical co-ordinates. Then we find that their conjugate momenta are, respectively, $p_i^\dagger = \Lambda_i = \beta_i\sqrt{(\mu_i)a_i}$, $p_{n+i}^\dagger = \Pi_i = \Lambda_i\left(\sqrt{(1-e_i^2)} - 1\right)$, and $p_{2n+i}^\dagger = \Lambda_i\sqrt{(1-e_i^2)}(\cos i_i - 1)$, the Hamiltonian function for these being

$$H^\dagger(q^\dagger, p^\dagger) = -\frac{1}{2}\sum_{i=1}^{n} \frac{\mu_i^3\beta_i^3}{\Lambda_i^2} - R$$

As before, we may write $R = \sum_j K_j \cos N_j$, with $j = (j_1, j_2, \ldots, j_{3n})$, and $N_j = \sum_{i=1}^{n}(j_i\lambda_i + j_{n+i}\varpi_i + j_{2n+i}\Omega_i)$, the summation being over all sets j with $j_1 \geq 0$ and $\sum_{i=1}^{n}(j_i + j_{n+i} + j_{2n+i}) = 0$. To avoid the singularities at $e_i = 0$ and $i_i = 0$ (actually *simple cases!*), we transform to rectangular-type parameters

$$\xi_i = \sqrt{(-2\Pi_i)}\cos\varpi_i = \sqrt{(\Lambda_i)}\varepsilon_i\cos\varpi_i,$$

$$\eta_i = \sqrt{(-2\Pi_i)}\sin\varpi_i = \sqrt{(\Lambda_i)}\varepsilon_i\sin\varpi_i,$$

and
$$q_i = 2\sqrt{(\Lambda_i + \Pi_i)}\gamma_i \cos \Omega_i,$$

$$p_i = 2\sqrt{(\Lambda_i + \Pi_i)}\gamma_i \sin \Omega_i, \text{ where}$$

$$\varepsilon_i = \sqrt{2}\sqrt{\left(1 - \sqrt{(1 - e_i^2)}\right)} = e_i + \tfrac{3}{32}e_i^3 + \tfrac{3}{16}e_i^5 + \ldots,$$

and $\gamma_i = \sin\left(\tfrac{1}{2}i_i\right)$. Then the (λ_i, ξ_i, q_i) are canonical co-ordinates, with their conjugate momenta (Λ_i, η_i, p_i), respectively, the Hamiltoniam function for these being

$$H^{\ddagger}(\lambda, \xi, q; \Lambda, \eta, p) = -\frac{1}{2}\sum_{i=1}^{n} \frac{\mu_i^3 \beta_i^3}{\Lambda_i^2} - R$$

with $R = \sum_j (B_j \cos N_j + C_j \sin N_j)$, in which $N_j = \sum_{i=1}^{n} j_i \lambda_i$, the sum being over sets $j = (j_1, j_2, \ldots, j_n)$ with $j_1 \geq 0$ and $\sum_{i=1}^{n} j_i = 0$. Here B_j and C_j are power series in the $\xi_i, \eta_i, q_i,$ and p_i, and from the d'Alembert property, the B_j are of even degree in the η_i and p_i, and the C_j of odd degree in these.

4.3 Separation of the short-period terms from the long-period problem

We will choose a generating function W to give us a transformation

$$(\lambda, \xi, q; \Lambda, \eta, p) \mapsto (\tilde{\lambda}, \tilde{\xi}, \tilde{q}; \tilde{\Lambda}, \tilde{\eta}, \tilde{p}) \tag{25}$$

defined by

$$\lambda_i = \tilde{\lambda}_i + \sum_{j=1}^{\infty} \frac{1}{j!}\mathcal{L}_W^j(\tilde{\lambda}_i)$$

$$\xi_i = \tilde{\xi}_i + \sum_{j=1}^{\infty} \frac{1}{j!}\mathcal{L}_W^j(\tilde{\xi}_i)$$

$$q_i = \tilde{q}_i + \sum_{j=1}^{\infty} \frac{1}{j!}\mathcal{L}_W^j(\tilde{q}_i)$$

$$\Lambda_i = \tilde{\Lambda}_i + \sum_{j=1}^{\infty} \frac{1}{j!}\mathcal{L}_W^j(\tilde{\Lambda}_i)$$

$$\eta_i = \tilde{\eta}_i + \sum_{j=1}^{\infty} \frac{1}{j!}\mathcal{L}_W^j(\tilde{\eta}_i)$$

$$p_i = \tilde{p}_i + \sum_{j=1}^{\infty} \frac{1}{j!}\mathcal{L}_W^j(\tilde{p}_i) \tag{26}$$

Since the equations giving the transformation have no explicit dependence on time, the new Hamiltonian function, $\tilde{H}$, is equal to the old, so

$$H^{\ddagger}(\lambda, \xi, q; \Lambda, \eta, p) = \tilde{H}(\tilde{\lambda}, \tilde{\xi}, \tilde{q}; \tilde{\Lambda}, \tilde{\eta}, \tilde{p})$$

$$= \tilde{H}(\lambda, \xi, q; \Lambda, \eta, p) + \sum_{j=1}^{\infty} \frac{1}{j!} \mathcal{L}_W^j(\tilde{H}) \qquad (27)$$

Write $H^{\ddagger} = H_0^{\ddagger} - R$, where $H_0^{\ddagger} = -\frac{1}{2} \sum_{i=1}^{n} \left(\mu_i^3 \beta_i^3 / \Lambda_i^2 \right)$. Now the disturbing function will be given by an expansion in powers of the perturbation parameter ϵ of the form $R = \sum_{j=1}^{\infty} R_j$, where R_j is of order ϵ^j, and we will find that the generating function, W, and the new Hamiltonian function, $\tilde{H}$, will both need to be expanded, so put $W = \sum_{j=1}^{\infty} W_j$, and $\tilde{H} = \sum_{j=1}^{\infty} \tilde{H}_j$, where W_j and $\tilde{H}_j$ are each of order ϵ^j. Now, in Eq. (27), we first equate terms independent of ϵ, giving $\tilde{H}_0 = H_0^{\ddagger}$, that is, the unperturbed motion is not changed by this transformation, as we expect. Then, equating terms of order ϵ, we get

$$\begin{aligned} \tilde{H}_1(\lambda, \xi, q; \Lambda, \eta, p) &= -R_1 + \{H_0^{\ddagger}, W_1\} \\ &= -R_1 - \sum_{i=1}^{n} n_i \frac{\partial W_1}{\partial \lambda_i} \end{aligned} \qquad (28)$$

since n_i is $\frac{\partial H_0^{\ddagger}}{\partial \Lambda_i}$. Then, for each term $B_j \cos \left(\sum_{i=1}^{n} j_i \lambda_i \right)$ in R_1, we include in W_1 the term $-B_j / \left(\sum_{i=1}^{n} j_i n_i \right) \sin \left(\sum_{i=1}^{n} j_i \lambda_i \right)$, and for each term $C_j \sin \left(\sum_{i=1}^{n} j_i \lambda_i \right)$ in R_1, we include in W_1 the term $+C_j / \left(\sum_{i=1}^{n} j_i n_i \right) \cos \left(\sum_{i=1}^{n} j_i \lambda_i \right)$. Then $\tilde{H}_1$ is left equal to those terms of $-R_1$ which are independent of the λ_i, that is without the short-period terms. We now equate terms in (27) which are of degree 2 in ϵ,

$$\begin{aligned} \tilde{H}_2(\lambda, \xi, q; \Lambda, \eta, p) &= -R_2 + \{H_0^{\ddagger}, W_2\} + \frac{1}{2}\{\{H_0^{\ddagger}, W_1\}, W_1\} \\ &= -R_2^* - \sum_{i=1}^{n} n_i \frac{\partial W_2}{\partial \lambda_i} \end{aligned}$$

where

$$R_2^* = -R_2 + \frac{1}{2}\{\{H_0^{\ddagger}, W_1\}, W_1\}$$

As in the choice of W_1, we choose W_2 to contain short-period terms, leaving $\tilde{H}_2$ with only long-period terms, independent of the λ_i. This process can continue indefinitely through successive orders of ϵ, with each W_j containing only short-period terms, and each $\tilde{H}_j$ having only long-period terms. Then the short-period terms in the perturbations may be read off from (26), since the $(\tilde{\lambda}_i, \tilde{\xi}_i, \tilde{q}_i, \tilde{\Lambda}_i, \tilde{\eta}_i, \tilde{p}_i)$ will contain only long-period perturbations.

4.4   Solution of the long-period part

The d'Alembert property implies that the terms in $\tilde{H}$ of lowest degree in the $(\tilde{\xi}_i, \tilde{q}_i, \tilde{\eta}_i, \tilde{p}_i)$ will be quadratic, and so of form

$$\sum_{j,k} \left(K_{j,k}(\tilde{\xi}_j \tilde{\xi}_k + \tilde{\eta}_j \tilde{\eta}_k) + K_{j,k}^*(\tilde{q}_j \tilde{q}_k + \tilde{p}_j \tilde{p}_k) \right) \qquad (29)$$

the summation being over all pairs (j, k) with $1 \geq j \geq k \geq n$, the coefficients $K_{j,k}$ and $K^*_{j,k}$ being functions of the Λ_i only. The *secular variations* theory of Laplace and Lagrange, developed in the eighteenth century, was equivalent to keeping just these quadratic terms, and omitting all higher terms. Therefore the equations of motion for the equivalent of $(\tilde{\xi}_i, \tilde{q}_i, \tilde{\eta}_i, \tilde{p}_i)$ in their theory were linear, and their solution is a superposition of oscillations, each represented by circular motions, each uniformly described, in the $(\tilde{\xi}_i, \tilde{\eta}_i)$ and $(\tilde{q}_i, \tilde{p}_i)$ planes, of periods between 47,000 and 2 million years. We proceed now to a linear transformation of the complete equations which reduces these quadratic terms to diagonal form, that is, to put the Laplace and Lagrange secular theory into normal coordinates, though we will apply it to the complete system: $(\tilde{\lambda}, \tilde{\xi}, \tilde{q}; \tilde{\Lambda}, \tilde{\eta}, \tilde{p}) \mapsto (\breve{\lambda}, \sigma; \breve{\Lambda}, \tau)$, given by

$$\sigma_r = \sum_{j=1}^{n} a_{r,j} \tilde{\xi}_j$$

$$\tau_r = \sum_{j=1}^{n} a_{r,j} \tilde{\eta}_j$$

$$\sigma_{n+r} = \sum_{j=1}^{n} b_{r,j} \tilde{q}_j$$

$$\tau_{n+r} = \sum_{j=1}^{n} b_{r,j} \tilde{p}_j$$

The condition for this transformation to be canonical is in fact that the matrices of the $a_{i,j}$ and $b_{i,j}$ be orthogonal. Hence the inverse is

$$\tilde{\xi}_j = \sum_{j=1}^{n} a_{r,j} \sigma_r$$

$$\tilde{\eta}_j = \sum_{j=1}^{n} a_{r,j} \tau_r$$

$$\tilde{q}_j = \sum_{j=1}^{n} b_{r,j} \sigma_{n+r}$$

$$\tilde{p}_j = \sum_{j=1}^{n} b_{r,j} \tau_{n+r}$$

We also note that $\breve{\Lambda}_i = \tilde{\Lambda}_i$ and that

$$\breve{\lambda}_i = \tilde{\lambda}_i + \sum_{r=1}^{n} \sum_{s=1}^{n} \sum_{k=1}^{n} \left(a_{r,k} \frac{\partial a_{s,k}}{\partial \Lambda_i} \sigma_r \sigma_s + b_{r,k} \frac{\partial b_{s,k}}{\partial \Lambda_i} \sigma_{n+r} \sigma_{n+s} \right) \tag{30}$$

The Hamiltonian function for the transformed system is $\breve{H}(\breve{\lambda}, \sigma; \breve{\Lambda}, \tau)$ which, because of the absence of explicit dependence on time in the equations for the transformation, is equal to $\tilde{H}(\tilde{\lambda}_i, \tilde{\xi}_i, \tilde{q}_i, \tilde{\Lambda}_i, \tilde{\eta}_i, \tilde{p}_i)$, and has the form

$$\check{H} = -\frac{1}{2} \sum_{i=1}^{n} \mu_i^2 \beta_i^3 / \check{\Lambda}_i - \check{R} \tag{31}$$

where

$$\check{R} = -\frac{1}{2} \sum_{r=1}^{2n} \gamma_r (\sigma_r^2 + \tau_r^2)$$

$+$ terms in the σ_r and τ_r of order 4, 6, 8, $\ldots$

here all the terms in $\check{R}$ are power series in the quantities of the types $\sigma_r^2 + \tau_r^2, \sigma_r \sigma_s + \tau_r \tau_s$, and the γ_r and other coefficients are functions of the $\check{\Lambda}_i$. The γ_r, for $1 \geq r \geq n$, are the eigenvalues of the matrix a, and the γ_r, for $n+1 \geq r \geq 2n$, are the eigenvalues of the matrix b, and these are the limiting values, as the amplitudes of the secular variations tend to zero, of the frequencies, in the usual secular variation theory, often denoted by "g_r" and "s_r". We also note that $\check{\lambda}_i = \tilde{\lambda}_i$, and $\check{\lambda}_i$ is an *ignorable* coordinate, that is, it does not appear explicitly in the Hamiltonian function, so that its conjugate momentum, $\check{\Lambda}_i$, is a constant of the motion. Now carry out a further transformation, to return the secular motions to action and angle form:

$$(\check{\lambda}, \sigma; \check{\Lambda}, \tau) \mapsto (\lambda^\dagger, \omega; \Lambda^\dagger, \Omega)$$

defined by

$$\sigma_r = \sqrt{(-2\Omega_r)} \cos \omega_r$$
$$\tau_r = \sqrt{(-2\Omega_r)} \sin \omega_r$$

for $r = 1, 2, \ldots, n$. Note that

$$\sigma_r \sigma_s + \tau_r \tau_s = 2\sqrt{(\Omega_r \Omega_s \cos(\omega_r - \omega_s))}$$

The Hamiltonian function now becomes

$$H^\dagger(\lambda^\dagger, \omega; \Lambda^\dagger, \Omega) = -\sum_{i=1}^{n} \mu_i^2 \beta_i^3 / (2\Lambda^{\dagger 2}) - R^\dagger \tag{32}$$

where $R^\dagger = -\sum_{r=1}^{2n} \gamma_r \Omega_r + \sum_\nu K_\nu^\dagger \cos P_\nu$ and $P_\nu = \sum_{r=1}^{2n} \nu_r \omega_r$, the summation in (32) being over those sets $\nu = (\nu_1, \nu_2, \ldots, \nu_{2n})$ with $\sum_{r=1}^{2n} \nu_r = 0$. Note that the d'Alembert property links the power of $\sqrt{(\Omega_r)}$, which appears as a factor in $K_\nu^\dagger$, to the multiple of the ω_r which appears in the corresponding angular argument P_ν. A final transformation:

$$(\lambda^\dagger, \omega; \Lambda^\dagger, \Omega) \mapsto (\lambda^*, \omega^*; \Lambda^*, \Omega^*) \tag{33}$$

similar in nature to the first one (25), removes the terms involving the ω_r, and so leaves the final Hamiltonian function, H^*, dependent only on the momenta Λ_r^* and Ω_r^*, which are therefore constants of the motion, and the angles λ_r^* and ω_r^* are linear functions of the time, their rates of change being $\frac{\partial H^*}{\partial \Lambda_r^*}$ and $\frac{\partial H^*}{\partial \Omega_r^*}$ respectively, and therefore constants. The resulting formal expressions for the original orbital parameters, obtained by proceeding backwards through the transformations, are therefore entirely composed of multiple periodic functions of the linear arguments λ_r^* and ω_r^*, indicating that all perturbations are periodic, though some of the periods are very long. For more details of this and the other transformations used here, see Message (1988).

4.5 Properties of the complete solution

The Laplace and Lagrange secular variations theory, confined to the terms quadratic in the eccentricities and orbital inclinations, indicated the stability of the solar system. We now see that when the theory is extended to any arbitrary power of these, the solution is still expressible entirely in periodic terms. Does this not complete the argument to establish the stability of the system? However, recall the denominators $\sum_{i=1}^{n} j_i n_i$, of which one can always be found of any arbitrary smallness, and which prevent the multiple Fourier series from being uniformly convergent in any finite piece of the space of solutions, denying us any analytical confirmation of the validity of the series. But such perturbation series have been very successful in predicting planetary positions over many centuries, and in successful comparison with the results of numerical integrations of the equations of motion, even over hundreds of thousands of years. To understand how this could be so, note that, a small value of the denominator $\sum_{i=1}^{n} j_i n_i$ will usually imply large values of some at least of the $|j_i|$ (recall that $\sum_{i=1}^{3n} j_i = 0$) and the d'Alembert property will therefore mean there is as a factor in the coefficient a large power of at least some of the e_i and $|\sin i_i|$, which will counteract the small denominator, in a system, such as the solar system, in which the orbits are nearly circular, and nearly coplanar. There is a large literature on the analytical aspects of this. See for example, Arnol'd (1963) and Whiteman (1977).

4.6 Small-integer commensurabilities

But if there are small-integer commensurabilities between the mean orbital motions of any two planets, say P_i and P_j, say

$$(p+q)n_i \approx pn_j \qquad (34)$$

where p and q are two small integers, then these considerations using the d'Alembert property do not apply, since, for example, the term in R' with argument

$$\Theta = (p+q)\lambda_i - p\lambda_j - q\varpi_i$$

will have coefficient with the factor e_i^q, which will not be small unless e_i is. If such a near-commensurability is close enough, then the series expressions for the perturbations, derived by the methods which we have been considering, using the transformation (25), may not even approximate to the actual behaviour of the system. We may call this the situation of *deep resonance*. The perturbations may then be of a quite different character, and different methods for solution of the equations are necessary. See for example, the cases which arise in the satellite system of the planet Saturn, as described, for example, in Message (1999). We may speak of *shallow resonance* in the case of an approximate resonance which is not so close that the usual methods fail, but where the small denominator nevertheless causes the corresponding term in the perturbation to be large, as, for example, in the mutual perturbations of the planets Jupiter and Saturn, whose orbital mean motions are nearly in the ratio 5:2. There are consequently large terms, especially in the mean longitudes of these planets (the "great inequality").

Likewise, of course, limitations on our solution arise in the case of near resonances within the frequencies, ω_r, arising in the "secular variations", which will lead to small denominators in the generating function of the final transformation (33). (Note, however, that although one of the frequencies has a zero value, corresponding to the solution of the equations in which the motion of the planetary system remains planar, this does not lead to difficulty, since there is no term in this transformation having this frequency alone as a denominator.)

But, nevertheless, Laskar—in his analysis of the results of his numerical integrations of the equations for the secular variations of the eight major planets (Laskar 1991)—finds, for example, a critical argument in the secular motion,

$$2\omega_4 - 2\omega_3 + \Omega_3 - \Omega_4$$

(in the notation we are using here), associated with the apses and nodes of the orbits of the Earth and Mars, which does not behave as our expressions would predict, but, over the 200 million years corresponding to his numerical investigations, makes transitions between different types of behaviour, from an interval during which it is librating about the value zero, through an interval during which it is changing monotonically through complete revolutions, to a further interval during which it is oscillating about the value 180 degrees.

References

Arnol'd, V.I. (1963), 'Small denominators and problems of stability of motion in classical and celestial Mechanics', *Usp. Mat. Nauk* **18**, 91-192 and *Russian Mathematical Surveys*, **18**, 85-193.

Hori, G. (1966), 'Theory of general perturbation with unspecified canonical variable', *Publications of the Astronomical Society of Japan* **18**, 287-296.

Laskar, J. (1991), *ICARUS* **88**, 266-291.

Message, P.J. (1966), 'On nearly-commensurable periods in the restricted problem of three bodies, with calculations of the long-period variations in the interior 2:1 case', in *Proceedings of International Astronomical Union Symposium No. 25*, 197-222.

Message, P.J. (1987), 'Planetary perturbation theory from Lie series, including resonance and critical arguments', in *Long-term dynamical behaviour of natural and artificial N-body systems*, ed. A.E. Roy, Kluwer, 47-72.

Message, P.J. (1988), 'Planetary perturbation theory from Lie series', *Celestial Mechanics*, **43**, 119-125.

Message, P.J. (1999), 'Orbits of Saturn's satellites: some aspects of commensurabilities and periodic orbits', in *The Dynamics of Small Bodies in the Solar System*, ed. B.A. Steves and A.E. Roy, Kluwer, 207-225.

Whiteman, K.J. (1977), 'Invariants and stability in classical mechanics', *Rep. Prog. Phys*, **40**, 1033-1069.

Fundamentals of regularization in celestial mechanics and linear perturbation theories

Jörg Waldvogel

Seminar for Applied Mathematics, Swiss Federal Institute of Technology
ETH, CH-8092 Zurich

What is regularization? Why is it useful for studying planetary systems? Although collisions in evolved planetary systems are rare events, the theory of regularization (i.e. the removal of collision singularities) is an important tool for developing good perturbation theories and efficient numerical algorithms. We give an overview of these techniques.

1 Introduction

Orbital mechanics of extra-solar planetary systems is based on the widely accepted and well-founded assumption that the Newtonian law of gravitation is also valid outside our solar system; in fact we postulate its *universal* validity. Hence many of the basic ingredients of exoplanet research are identical with the foundations of classical celestial mechanics: the motion of a point mass under the Newtonian attraction of a central body and a small perturbation. This fundamental topic, called the perturbed *problem of two bodies*, or for short, the *perturbed Kepler problem* is the main interest of this contribution.

It is therefore appropriate to include here this chapter on classical celestial mechanics in order to revisit the perturbed Kepler problem and its solution. Our goal is to recover the classical theory of Kepler motion and to present a simple theory of the perturbed Kepler problem. For classical perturbation theories see, e.g., Brouwer and Clemence (1961). The common tool is *regularization*, a transformation of both space and time variables introduced by Levi-Civita (1920) in the plane and generalized by Kustaanheimo and Stiefel (1965) in space. Historically, regularization was developed for investigating the singularities of Kelper motion and for describing collisions of two point masses as well as for improving the numerical integration of (near-)collision orbits.

In this context, however, a mere side-effect of the above-mentioned regularizations will be exploited: the *linearity* of the transformed differential equations of motion. Thanks to this linearity and by taking advantage of a unified theory of the planar and spatial cases we will be able to present an elegant treatment of the basics of orbital mechanics.

We illustrate the simplicity of handling perturbed linear problems by means of the following simple example. Consider, e.g., the perturbed system

$$\dot{x}(t) + A(t)\,x(t) - b(t) = \varepsilon f(x,t), \quad x : t \in R \mapsto x(t) \in R^n, \qquad (1)$$

of linear differential equations, where $A(t)$ is a given time-dependent $(n \times n)$-matrix. Eq. (1) may formally be solved to arbitrary order by the series

$$x(t) = x_0(t) + \varepsilon\,x_1(t) + \varepsilon^2\,x_2(t) + \dots ,$$

where $x_k(t)$ satisfies the linear differential equation

$$\dot{x}_k(t) + A(t)\,x_k(t) = f_{k-1}(t), \quad k = 0, 1, 2, \dots . \qquad (2)$$

Here $f_{-1}(t) := b(t)$, and $f_0(t), f_1(t), \dots$ are defined as the coefficients of the formal Taylor series of $f(x,t)$ with respect to ε:

$$\sum_{k=0}^{\infty} \varepsilon^k\,f_k(t) = f\big(x_0(t) + \varepsilon\,x_1(t) + \varepsilon^2\,x_2(t) + \dots , t\big) .$$

Note that the linear differential equations (2) are all of the type of the unperturbed problem $k = 0$; they only differ in their right-hand sides.

We will begin by collecting the well-known classical formulas governing planar elliptic Kepler motion (essentially the three Keplerian laws and some geometry of conic sections). Simple properties of these equations will provide a natural motivation for Levi-Civita's regularization. In Section 3 we will describe in detail the planar regularization procedure and show that the resulting linear system of differential equations is even of a very special form: a harmonic oscillator. Section 4 will present the corresponding spacial regularization, elegantly represented by means of quaternions. In Section 5 we will discuss the linear perturbation theories of Kepler motion as a perturbed harmonic oscillator.

2   Planar Kepler motion

Consider Kepler motion of a massless particle positioned at $x = \binom{x_1}{x_2} \in R^2$ around a central body with gravitational parameter μ. For convenience the position of the particle will equivalently be denoted by the complex coordinate

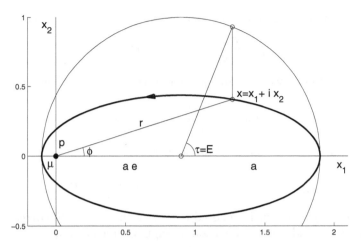

Figure 1. *The planar elliptic Kepler motion with eccentricity* $e = 0.9$

$\mathbf{x} = x_1 + i\,x_2 \in C$ (complex numbers will be denoted by boldface characters). Kepler motion is then governed by the differential equation

$$\ddot{\mathbf{x}} + \mu\,\frac{\mathbf{x}}{r^3} = 0\,, \qquad r = |\,\mathbf{x}\,|\,, \tag{3}$$

where dots denote derivatives with respect to time t. The Keplerian orbit of Figure 1 is standardized such that the apocenter is on the x_1-axis; therefore the orbit is given by merely two *orbital elements*, e.g. the major semi-axis a and the eccentricity e. An important orbital element depending on a and e is the semi-latus rectum $p = a\,(1 - e^2)$. Quantities varying with the moving particle are the radial distance $r = |\,\mathbf{x}\,|$ and the polar angle $\varphi = \arg(\mathbf{x})$ as well as the eccentric anomaly E (see Figure 1), which turns out to be the parameter best suited for completely describing Kepler motion in space and time.

The famous relations describing the orbit are

$$x_1 = a\,(e + \cos E)\,, \qquad x_2 = a\,\sqrt{1 - e^2}\cdot\sin E\,, \tag{4}$$

which implies

$$r = |\,\mathbf{x}\,| = a\,(1 + e\,\cos E)\,. \tag{5}$$

For determining the time t (normalized such that a passage through the apocenter is at $t = 0$) we may use the famous *Keplerian equation* and its derivative,

$$t = \sqrt{\frac{a^3}{\mu}}\cdot(E + e\,\sin E)\,, \qquad \frac{dt}{dE} = \sqrt{\frac{a}{\mu}}\cdot r\,. \tag{6}$$

Keplerian orbits have a simple representation in polar coordinates,

$$r = \frac{p}{1 - e \cos \varphi}, \qquad p = a\left(1 - e^2\right), \tag{7}$$

where the relation between E and φ, precise mod 2π, may be written as

$$\tan(\frac{\varphi}{2}) = \sqrt{\frac{1 - e}{1 + e}} \, \tan(\frac{E}{2}). \tag{8}$$

Finally, we mention the conservation of energy

$$\frac{1}{2} \, |\dot{\mathbf{x}}|^2 - \frac{\mu}{r} = -h = \text{const}, \quad h = \frac{\mu}{2\,a} > 0, \tag{9}$$

where the energy constant is denoted by $-h$, such that $h > 0$ corresponds to the elliptic case, and the conservation of angular momentum,

$$|\,x \times \dot{x}\,| = \sqrt{\mu\,p} = \text{const}. \tag{10}$$

We now try to exploit properties of these relations in order to find appropriate variables for a simple description of Kepler motion.

(i) Equations (4) and (6) suggest that E might be a more suitable independent variable than the time t. In generalization of the transformation used by Sundman (1907) we introduce a *fictitious time* τ according to the differential relation

$$dt = \frac{r}{c} \cdot d\tau, \quad r = |\mathbf{x}|. \tag{11}$$

With $c = 1$, τ is Sundman's variable, whereas, according to Eqs. (6) and (9), $c = \sqrt{2\,h}$ introduces the eccentric anomaly $\tau = E$ as the new independent variable.

(ii) Consider the complex position $\mathbf{x} \in C$, written in terms of E by means of Eq. (4),

$$\mathbf{x} = x_1 + i\,x_2 = a \left(e + \cos E + i \sqrt{1 - e^2} \, \sin E \right).$$

Note that $\mathbf{x} = \mathbf{u}^2 \in C$ is identically the square of

$$\mathbf{u} = \sqrt{a\,(1 + e)} \, \cos(\frac{E}{2}) + i \sqrt{a\,(1 - e)} \, \sin(\frac{E}{2}). \tag{12}$$

This is precisely the conformal mapping between the parametric $\mathbf{u}$-plane and the physical $\mathbf{x}$-plane used by Levi-Civita: a conformal *squaring*.

3   The Levi-Civita transformation

In this section we carry out the *three* steps necessary for regularizing the unperturbed planar Kepler problem by Levi-Civita's method. The first two steps exactly follow the suggestions (i) and (ii) discussed above. The third step will merely consist of fixing the energy. In the following, we will therefore subject the equations of motion (3) as well as the energy equation (9) to the transformations of the first two steps. Complex notation will be used throughout, i.e. instead of the vector $x = \binom{x_1}{x_2} \in R^2$ we use the corresponding complex coordinate $\mathbf{x} = x_1 + i\,x_2 \in C$.

3.1 First step: slow-motion movie

Instead of the physical time t a new independent variable τ, the fictitious time, is introduced by the differential relation (11); derivatives with respect to τ will be denoted by primes. Therefore the ratio $dt/d\tau$ of two infinitesimal increments is made proportional to the distance r; the movie is run in slow-motion whenever r becomes small. With the differentiation rules

$$\frac{d}{dt} = c\,r^{-1}\frac{d}{d\tau}\,, \qquad \frac{d^2}{dt^2} = c^2\left(r^{-2}\frac{d^2}{d\tau^2} + (\frac{c'}{c}r - r')\,r^{-3}\frac{d}{d\tau}\right)$$

Eqs. (3) and (9) are transformed into

$$c^2\left(r\,\mathbf{x}'' + (\frac{c'}{c}r - r')\,\mathbf{x}'\right) + \mu\,\mathbf{x} = 0\,, \qquad \frac{1}{2}c^2 r^{-2}\,|\mathbf{x}'|^2 - \frac{\mu}{r} = -h\,. \qquad (13)$$

3.2 Second step: conformal squaring

This part of Levi-Civita's regularization procedure consists of representing the complex physical coordinate $\mathbf{x}$ as the square $\mathbf{u}^2$ of a complex variable $\mathbf{u} = u_1 + i\,u_2 \in C$,

$$\mathbf{x} = \mathbf{u}^2\,, \qquad (14)$$

i.e. the mapping from the parametric plane to the physical plane is chosen as a conformal squaring. Eq. (14) implies

$$r = |\mathbf{x}| = |\mathbf{u}|^2 = \mathbf{u}\bar{\mathbf{u}}\,, \qquad (15)$$

and differentiation of Eqs. (14) and (15) yields

$$\mathbf{x}' = 2\,\mathbf{u}\mathbf{u}'\,, \qquad \mathbf{x}'' = 2\left(\mathbf{u}\mathbf{u}'' + \mathbf{u}'^{\,2}\right) \in C\,, \qquad r' = \mathbf{u}'\bar{\mathbf{u}} + \bar{\mathbf{u}}'\,. \qquad (16)$$

Substitution of this into Eq. (13) and cancelling two equal terms ($2\,r\,\mathbf{u}'^2$ and $2\,\mathbf{u}'\,\bar{\mathbf{u}}\,\mathbf{u}\,\mathbf{u}'$) as well as dividing by $c^2\,\mathbf{u}$ yields

$$2\,r\,\mathbf{u}'' + 2\,\frac{c'}{c}\,r\,\mathbf{u}' + (\frac{\mu}{c^2} - 2\,|\mathbf{u}'|^2)\,\mathbf{u} = 0\,. \qquad (17)$$

Although we have $c' = 0$ in this section, the second term has been retained in view of the perturbed Kepler motion to be considered later. Similarly, substitution into the energy equation (13) yields

$$2\,c^2\,|\mathbf{u}'|^2 = \mu - r\,h\,. \qquad (18)$$

3.3 Third step: fixing the energy

This final regularization step amounts to considering only orbits of the fixed energy $-h$. Then, the nonlinear factor $|\mathbf{u}'|^2$ may conveniently be eliminated from the Eqs. (17) and (18). Multiplying the result by c^2/r yields

$$2\,c^2\,\mathbf{u}'' + 2\,c\,c'\,\mathbf{u}' + h\,\mathbf{u} = 0, \quad \mathbf{u} \in C,. \tag{19}$$

In the unperturbed case $c' = 0$ this reduces to the differential equation

$$\mathbf{u}'' + \omega^2\,\mathbf{u} = 0, \quad \mathbf{u} \in C, \quad \omega := c^{-1}\sqrt{h/2} \tag{20}$$

of a harmonic oscillator of frequency ω in 2 dimensions. For Sundmun's choice $c = 1$ Eq. (20) simply becomes $2\,\mathbf{u}'' + h\,\mathbf{u} = 0$.

Remark 1. Based on the preceding part of this section we propose an alternate way of deriving the Kepler formulas of Section 2. We depart from the equations of motion (3), possibly together with the energy integral (9) and the angular-momentum integral (10). Next, carrying out the three steps of the Levi-Civita regularization procedure results in Eq. (19). A favourable choice is $c = \sqrt{2\,h} = $ const, resulting in the oscillator equation $\mathbf{u}'' + \frac{1}{4}\,\mathbf{u} = 0$. With no loss of generality (and in an appropriate normalization) we write its general solution as $\mathbf{u} = A\,\cos(\frac{E}{2}) + i\,B\,\sin(\frac{E}{2})$ with $A, B \in R$.

Now the entire theory of planar Kepler motion may be recovered from elementary calculations and some geometry of conic sections:

- the parametrization (4) and Eq. (5)

- the geometric meaning of E

- the energy integral and Eq. (9)

- the time evolution, Kepler's equation, Eq. (6)

- Kepler's third law

- the orbit in polar coordinates, Eq. (7)

- the angular-momentum integral, Eq. (10)

Remark 2. Obtaining initial values $\mathbf{u}(0) = \sqrt{\mathbf{x}(0)} \in C$ requires the computation of a complex square root. This can conveniently be accomplished by means of the formula

$$\sqrt{\mathbf{x}} = \frac{\mathbf{x} + |\mathbf{x}|}{\sqrt{2\,(|\mathbf{x}| + \mathrm{Re}\,\mathbf{x})}}, \tag{21}$$

which reflects the observation that the complex vector $\sqrt{\mathbf{x}}$ has the direction of the bisector between $\mathbf{x}$ and the real vector $|\mathbf{x}|$; it holds in the range $-\pi < \arg(\mathbf{x}) < \pi$. The alternative formula

$$\sqrt{\mathbf{x}} = \frac{\mathbf{x} - |\mathbf{x}|}{i\sqrt{2\left(|\mathbf{x}| - \operatorname{Re}\mathbf{x}\right)}}$$

holds in $0 < \arg(\mathbf{x}) < 2\pi$ and agrees with (21) in the upper half-plane; it therefore provides the analytic continuation of (21) into the sector $\pi \leq \arg(\mathbf{x}) < 2\pi$. Furthermore, it avoids a loss of accuracy near the negative real axis $\mathbf{x} < 0$.

4 Spatial regularization with quaternions

In this section we indicate how Levi-Civita's regularization procedure may be generalized to three-dimensional motion. The essential step is to replace the conformal squaring of Section 3 by the Kustaanheimo-Stiefel (KS) transformation. A preliminary version of this transformation using spinor notation was proposed by Kustaanheimo (1964); the full theory was developed in a subsequent joint paper (Kustaanheimo and Stiefel, 1965); the entire topic is extensively discussed in the comprehensive text by Stiefel and Scheifele (1971). The relevant mapping from the 3-sphere onto the 2-sphere was already discovered by Heinz Hopf (1931) and is referred to in topology as the Hopf mapping.

Both the Levi-Civita and the Kustaanheimo-Stiefel regularization share the property of "linearizing" the equations of motion of the two-body problem; both are therefore well suited for developing linear perturbation theories. Quaternion algebra, introduced by W. R. Hamilton (1844), turns out to be the ideal tool for regularizing the three-dimensional Kepler motion, as was observed by M. D. Vivarelli (1983) and J. Vrbik (1994, 1995). It was observed by Waldvogel (2006a, 2006b) that by introducing an unconventional conjugation, the *star conjugation* of quaternions, (see the definition in Eq. (29) below) the spatial regularization procedure according to Kustaanheimo and Stiefel is in complete formal agreement with Levi-Civita's planar procedure described in Section 3. Here we will repeat or summarize the relevant parts of those papers.

4.1 Basics

Quaternion algebra is a generalization of the algebra of complex numbers obtained by using three independent "imaginary" units i, j, k. As for the single imaginary unit i in the algebra of complex numbers, the rules

$$i^2 = j^2 = k^2 = -1$$

are postulated, together with the non-commutative multiplication rules

$$ij = -ji = k, \quad jk = -kj = i, \quad ki = -ik = j.$$

Given the real numbers $u_l \in R$, $l = 0, 1, 2, 3$, the object

$$\mathbf{u} = u_0 + i\,u_1 + j\,u_2 + k\,u_3 \tag{22}$$

is called a *quaternion* $\mathbf{u} \in U$, where U denotes the set of all quaternions (in the remaining sections boldface characters denote quaternions). The sum $iu_1 + ju_2 + ku_3$ is called the *quaternion part* of $\mathbf{u}$, whereas u_0 is naturally referred to as its real part. The above multiplication rules and vector space addition define the *quaternion algebra*. Multiplication is generally non-commutative; however, any quaternion commutes with a real:

$$c\,\mathbf{u} = \mathbf{u}\,c, \quad c \in R, \ \mathbf{u} \in R, \tag{23}$$

and for any three quaternions $\mathbf{u}, \mathbf{v}, \mathbf{w} \in U$ the associative law holds:

$$(\mathbf{u}\,\mathbf{v})\,\mathbf{w} = \mathbf{u}\,(\mathbf{v}\,\mathbf{w}). \tag{24}$$

The quaternion $\mathbf{u}$ may naturally be associated with the corresponding vector $u = (u_0, u_1, u_2, u_3) \in R^4$. For later reference we introduce notation for 3-vectors in two important particular cases: $\vec{u} = (u_1, u_2, u_3) \in R^3$ for the vector associated with the *pure quaternion* $\mathbf{u} = i\,u_1 + j\,u_2 + k\,u_3$, and $\underline{u} = (u_0, u_1, u_2)$ for the vector associated with the quaternion with a vanishing k-component, $\mathbf{u} = u_0 + i\,u_1 + j\,u_2$.

For convenience we also introduce the vector $\vec{\imath} = (i, j, k)$; the quaternion $\mathbf{u}$ may then be written formally as $\mathbf{u} = u_0 + \langle \vec{\imath}, \vec{u} \rangle$, where the notation $\langle \cdot, \cdot \rangle$ refers to the dot product of vectors. For the two quaternion products of $\mathbf{u}$ and $\mathbf{v} = v_0 + \langle \vec{\imath}, \vec{v} \rangle$ we then obtain the concise expressions

$$\begin{aligned}
\mathbf{u}\,\mathbf{v} &= u_0 v_0 - \langle \vec{u}, \vec{v} \rangle + \langle \vec{\imath},\ u_0\,\vec{v} + v_0\,\vec{u} + \vec{u} \times \vec{v} \rangle \\
\mathbf{v}\,\mathbf{u} &= u_0 v_0 - \langle \vec{u}, \vec{v} \rangle + \langle \vec{\imath},\ u_0\,\vec{v} + v_0\,\vec{u} - \vec{u} \times \vec{v} \rangle,
\end{aligned} \tag{25}$$

where $\times$ denotes the vector product. Note that the non-commutativity shows only in the sign of the term with the vector product.

The *conjugate* $\bar{\mathbf{u}}$ of the quaternion $\mathbf{u}$ is defined as

$$\bar{\mathbf{u}} = u_0 - i\,u_1 - j\,u_2 - k\,u_3; \tag{26}$$

then the *modulus* $|\mathbf{u}|$ of $\mathbf{u}$ is obtained from

$$|\mathbf{u}|^2 = \mathbf{u}\,\bar{\mathbf{u}} = \bar{\mathbf{u}}\,\mathbf{u} = \sum_{l=0}^{3} u_l^2. \tag{27}$$

As transposition of a product of matrices, conjugation of a quaternion product reverses the order of its factors:

$$\overline{\mathbf{u}\,\mathbf{v}} = \bar{\mathbf{v}}\,\bar{\mathbf{u}}. \tag{28}$$

4.2 The KS map in the language of quaternions

We first revisit the KS transformation by using quaternion algebra and the unconventional "conjugate" $\mathbf{u}^\star$, referred to as the *star conjugate* of the quaternion $\mathbf{u} = u_0 + i\,u_1 + j\,u_2 + k\,u_3$,

$$\mathbf{u}^\star := u_0 + i\,u_1 + j\,u_2 - k\,u_3 \,, \tag{29}$$

introduced by Waldvogel (2006a). The star conjugate of $\mathbf{u}$ may be expressed in terms of the conventional conjugate $\bar{\mathbf{u}}$ as

$$\mathbf{u}^\star = k\,\bar{\mathbf{u}}\,k^{-1} = -k\,\bar{\mathbf{u}}\,k \,;$$

however, it turns out that the definition (29) leads to a particularly elegant treatment of KS regularization. The following elementary properties are easily verified:

$$\begin{aligned} (\mathbf{u}^\star)^\star &= \mathbf{u} \\ |\,\mathbf{u}^\star|^2 &= |\mathbf{u}|^2 \\ (\mathbf{u}\,\mathbf{v})^\star &= \mathbf{v}^\star\,\mathbf{u}^\star \,. \end{aligned} \tag{30}$$

Consider now the mapping

$$\mathbf{u} \in U \;\longmapsto\; \mathbf{x} = \mathbf{u}\,\mathbf{u}^\star \,. \tag{31}$$

Star conjugation immediately yields $\mathbf{x}^\star = (\mathbf{u}^\star)^\star\,\mathbf{u}^\star = \mathbf{x}$; hence $\mathbf{x}$ is a quaternion of the form $\mathbf{x} = x_0 + i\,x_1 + j\,x_2$ which may be associated with the vector $\underline{x} = (x_0, x_1, x_2) \in R^3$. From $\mathbf{u} = u_0 + i\,u_1 + j\,u_2 + k\,u_3$ we obtain

$$\begin{aligned} x_0 &= u_0^2 - u_1^2 - u_2^2 + u_3^2 \\ x_1 &= 2(u_0\,u_1 - u_2\,u_3) \\ x_2 &= 2(u_0\,u_2 + u_1\,u_3) \,, \end{aligned} \tag{32}$$

which is exactly the KS transformation in its classical form or – up to a permutation of the indices – the Hopf map. Therefore we have

Theorem 1: The KS transformation which maps $u = (u_0, u_1, u_2, u_3) \in R^4$ to $\underline{x} = (x_0, x_1, x_2) \in R^3$ is given by the quaternion relation

$$\mathbf{x} = \mathbf{u}\,\mathbf{u}^\star \,,$$

where $\mathbf{u} = u_0 + iu_1 + ju_2 + ku_3, \mathbf{x} = x_0 + ix_1 + jx_2$.

Corollary 1: The norms of the vectors $\underline{x}$ and u satisfy

$$r := \|\underline{x}\| = \|u\|^2 = \mathbf{u}\,\bar{\mathbf{u}} \,. \tag{33}$$

Proof: By appropriately combining the two conjugations and using the rules (23), (24), (27), (28), (30) we obtain

$$\|\underline{x}\|^2 = \mathbf{x}\,\bar{\mathbf{x}} = \mathbf{u}\,(\mathbf{u}^\star\,\bar{\mathbf{u}}^\star)\,\bar{\mathbf{u}} = |\mathbf{u}^\star|^2\,|\mathbf{u}|^2 = |\mathbf{u}|^4 = \|u\|^4 \,,$$

from where the statement follows.

4.3   The inverse map

Since the mapping (32) does not preserve the dimension, its inverse in the usual sense does not exist. However, the present quaternion formalism yields an elegant way of finding the corresponding *fibration* of the original space R^4. Being given a quaternion $\mathbf{x} = x_0 + i\,x_1 + j\,x_2$ with vanishing k-component, $\mathbf{x} = \mathbf{x}^*$, we want to find all quaternions $\mathbf{u}$ such that $\mathbf{u}\,\mathbf{u}^* = \mathbf{x}$. We propose the following solution in two steps:

First step: Find a particular solution $\mathbf{u} = \mathbf{v} = \mathbf{v}^* = v_0 + i\,v_1 + j\,v_2$ which has also a vanishing k-component. Since $\mathbf{v}\,\mathbf{v}^* = \mathbf{v}^2$ we may use Eq. (21), which was developed for the complex square root, also for the square root of a quaternion:

$$\mathbf{v} = \frac{\mathbf{x} + |\mathbf{x}|}{\sqrt{2\,(|\mathbf{x}| + x_0)}}\;.$$

Clearly, $\mathbf{v}$ has a vanishing k-component. *Second step:* The entire family of solutions (the fibre corresponding to $\mathbf{x}$, geometrically a circle in R^4 parametrized by the angle ϑ), is given by

$$\mathbf{u} = \mathbf{v} \cdot e^{k\,\vartheta} = \mathbf{v}\,(\cos\vartheta + k\sin\vartheta)\,.$$

Proof. $\mathbf{u}\,\mathbf{u}^* = \mathbf{v}\,e^{k\,\vartheta}\,e^{-k\,\vartheta}\,\mathbf{v}^* = \mathbf{v}\,\mathbf{v}^* = \mathbf{x}$

4.4   The regularization procedure with quaternions

In order to regularize the perturbed three-dimensional Kepler motion by means of the KS transformation it is necessary to look at the properties of the map (31) under differentiation.

The transformation (31) or (32) is a mapping from R^4 to R^3; it therefore leaves one degree of freedom in the parametric space undetermined. In KS theory (Kustaanheimo and Stiefel, 1965; Stiefel and Scheifele, 1971), this freedom is taken advantage of by trying to inherit as much as possible of the conformality properties of the Levi-Civita map, but other approaches exist (e.g., Vrbik 1995). By imposing the "bilinear relation"

$$2\,(u_3\,du_0 - u_2\,du_1 + u_1\,du_2 - u_0\,du_3) = 0 \tag{34}$$

between the vector $u = (u_0, u_1, u_2, u_3)$ and its differential du on orbits the tangential map of (32) becomes a linear map with an orthogonal (but non-normalized) matrix.

This property has a simple consequence on the differentiation of the quaternion representation (31) of the KS transformation. Considering the noncommutativity of the quaternion product, the differential of Eq. (31) becomes

$$d\mathbf{x} = d\mathbf{u} \cdot \mathbf{u}^* + \mathbf{u} \cdot d\mathbf{u}^*\,, \tag{35}$$

whereas (34) takes the form of a commutator relation,

$$\mathbf{u} \cdot d\mathbf{u}^\star - d\mathbf{u} \cdot \mathbf{u}^\star = 0 \,. \tag{36}$$

Combining (35) with the relation (36) yields the elegant result

$$d\mathbf{x} = 2\,\mathbf{u} \cdot d\mathbf{u}^\star \,, \tag{37}$$

i.e. the bilinear relation (34) of KS theory is equivalent with the requirement that the tangential map of $\mathbf{u} \mapsto \mathbf{u}\,\mathbf{u}^\star$ behaves as in a commutative algebra.

By using the tools collected in this section together with the differentiation rule (37) the regularization procedure outlined in Section 3 will now be carried out for the three-dimensional Kepler problem. Care must be taken to preserve the order of the factors in quaternion products. Exchanging two factors is permitted if one of the factors is real or if the factors are mutually conjugate. An important tool for simplifying expressions is regrouping factors of multiple products according to the associative law (24). In order to display the simplicity of this approach we present all the details of the formal computations.

(a) First step in space: slow-motion movie

Let $\mathbf{x} = x_0 + i\,x_1 + j\,x_2 \in U$ be the quaternion associated with the position vector $\underline{x} = (x_0, x_1, x_2) \in R^3$; then the unperturbed Kepler problem (3) in space, written in quaternion notation, is given by

$$\ddot{\mathbf{x}} + \mu\,\frac{\mathbf{x}}{r^3} = 0 \in U \,, \quad r = |\mathbf{x}| \,. \tag{38}$$

The first transformation step calls for introducing the fictitious time τ according to Eq. (11). We restrict ourselves to Sundman's choice $c = 1$, hence $dt = r \cdot d\tau$. The results are formally identical with Eq. (13) with $c = 1$,

$$r\,\mathbf{x}'' - r'\,\mathbf{x}' + \mu\,\mathbf{x} = 0 \,, \qquad \frac{1}{2\,r^2}\,|\mathbf{x}'|^2 - \frac{\mu}{r} = -h \,. \tag{39}$$

(b) Second step: KS transformation with quaternions

Instead of the conformal squaring according to Eq. (14) we use the KS transformation (31),

$$\mathbf{x} = \mathbf{u}\,\mathbf{u}^\star \,, \qquad r := |\mathbf{x}| = \mathbf{u}\,\bar{\mathbf{u}} \,. \tag{40}$$

Differentiation by means of the commutator relation (36) yields

$$\mathbf{x}' = 2\,\mathbf{u}\,\mathbf{u}^{\star'} \,, \quad \mathbf{x}'' = 2\,\mathbf{u}\,\mathbf{u}^{\star''} + 2\,\mathbf{u}'\,\mathbf{u}^{\star'} \,, \quad r' = \mathbf{u}'\,\bar{\mathbf{u}} + \mathbf{u}\,\bar{\mathbf{u}}' \,. \tag{41}$$

Substitution of (40) and (41) into (39) results in the lengthy equation

$$(\mathbf{u}\,\bar{\mathbf{u}})\,(2\,\mathbf{u}\,\mathbf{u}^{\star''} + 2\,\mathbf{u}'\,\mathbf{u}^{\star'}) - (\mathbf{u}'\,\bar{\mathbf{u}} + \mathbf{u}\,\bar{\mathbf{u}}')\,2\,\mathbf{u}\,\mathbf{u}^{\star'} + \mu\,\mathbf{u}\,\mathbf{u}^\star = 0 \,, \tag{42}$$

which is considerably simplified by observing that the second and third term
– after applying the distributive law – compensate:

$$2\,(\mathbf{u}\,\bar{\mathbf{u}})\,\mathbf{u}'\,\mathbf{u}^{*'} \; - \; 2\,\mathbf{u}'\,(\bar{\mathbf{u}}\,\mathbf{u})\,\mathbf{u}^{*'} \; = \; 0\,.$$

Furthermore, by means of (23), (24) and (36) the fourth term of (42) may be
simplified as follows:

$$-2\,(\mathbf{u}\,\bar{\mathbf{u}}')\,(\mathbf{u}\,\mathbf{u}^{*'}) \; = \; -2\,\mathbf{u}\,(\bar{\mathbf{u}}'\,\mathbf{u}')\,\mathbf{u}^{*} \; = \; -2\,|\,\mathbf{u}'\,|^2\,\mathbf{u}\,\mathbf{u}^{*}\,.$$

By using this and left-dividing by $\mathbf{u}$ Eq. (42) now becomes

$$2\,r\,\mathbf{u}^{*''} + (\,\mu - 2\,|\mathbf{u}'|^{\,2}\,)\,\mathbf{u}^{*} = 0\,, \tag{43}$$

in almost perfect formal agreement with Eq. (17) (with $c = 1$) of the planar
case.

(c) Third step: fixing the energy in space

From (41), (30), (33) we have

$$|\,\mathbf{x}'\,|^2 = \mathbf{x}'\,\bar{\mathbf{x}}' = 4\,\mathbf{u}\,(\,\mathbf{u}^{*'}\,\bar{\mathbf{u}}^{*'}\,)\,\bar{\mathbf{u}} = 4\,r\,|\,\mathbf{u}'\,|^{\,2}\,; \tag{44}$$

therefore Eq. (39) becomes

$$\mu - 2\,|\,\mathbf{u}'\,|^2 \; = \; r\,h \tag{45}$$

in formal agreement with Eq. (18) found for the planar case. Substituting this
into the star-conjugate of (43) and dividing by r yields the elegant final result

$$2\,\mathbf{u}'' + h\,\mathbf{u} = 0\,, \tag{46}$$

a differential equation in perfect agreement with the result found in the planar
case.

5   The perturbed spatial Kepler problem

We now consider the perturbed spatial Kepler problem,

$$\ddot{\mathbf{x}} + \mu\,\frac{\mathbf{x}}{r^3} = \varepsilon\,\mathbf{f}(\mathbf{x}, t)\,, \quad r = |\,\mathbf{x}\,|\,, \tag{47}$$

written in quaternion notation. $\mathbf{f}(\mathbf{x}, t)$ is the perturbing function, $\mathbf{x} \in U$ and
$\mathbf{f} \in U$ are quaternions with vanishing k-components, and ε is a small parameter. Note that in the perturbed case an energy equation formally identical
with (9) still holds. However, $h = h(t)$ and $a = a(t)$ are now slowly varying
functions of time, $a(t)$ being the *osculating* major semi-axis; $h(t)$ satisfies the
differential equaion

$$\dot{h} = -\langle\,\dot{x},\,\varepsilon\,f\,\rangle \quad \text{or} \quad h' = -\langle\,x',\,\varepsilon\,f\,\rangle\,, \tag{48}$$

where $\langle \cdot, \cdot \rangle$ denotes the dot product of 3-vectors.

In the following, we report the results of the regularization procedure outlined in Section 4. The details are left to the reader as an exercise. Step 1 with $c = 1$ yields

$$r\,\mathbf{x}'' - r'\,\mathbf{x}' + \mu\,\mathbf{x} = r^3\,\varepsilon\,\mathbf{f}(\mathbf{x}, t)$$

instead of Eq. (39). By using (44) the energy equation (39) again becomes

$$\mu - 2\,|\mathbf{u}'|^2 = r\,h.$$

The right-hand side of Eq. (42) becomes

$$\mathbf{u}\,\bar{\mathbf{u}}\,r^2\,\varepsilon\,\mathbf{f}(\mathbf{x}, t)$$

instead of 0. Simplification as in Section 4 as well as left-multiplication by $r^{-1}\,\mathbf{u}^{-1}$ and star conjugation finally yields the perturbing equation for the quaternion coordinate $\mathbf{u}$:

Theorem 2: KS regularization, as formulated in terms of quaternions in Section 4, transforms the perturbed Kepler problem (47) into the perturbed harmonic oscillator

$$2\,\mathbf{u}'' + h\,\mathbf{u} = r\,\varepsilon\,\mathbf{f}(\mathbf{x}, t)\,\bar{\mathbf{u}}^\star, \quad r = |\mathbf{u}|^2,$$

where $h = r^{-1}\,(\mu - 2\,|\mathbf{u}'|^2)$ is the negative of the (slowly varying) energy.

In the following summary we collect the complete set of differential equations defining the regularized system equivalent to the perturbed spatial Kepler problem (47). The harmonic oscillator of Theorem 2 appears in the first line. For stating an initial-value problem a starting value of $\mathbf{u}$ needs to be chosen according to Section 4.3. The corresponding initial velocity is obtained by solving (37) for $d\mathbf{u}$:

$$\frac{d\mathbf{u}}{d\tau} = \frac{1}{2r}\,\frac{d\mathbf{x}}{dt}\,\bar{\mathbf{u}}^\star.$$

Summary. Regularized system for the 3D perturbed Kepler problem (47).

$$
\begin{aligned}
2\,\mathbf{u}'' + h\,\mathbf{u} &= r\,\varepsilon\,\mathbf{f}(\mathbf{x}, t)\,\bar{\mathbf{u}}^\star, & r &= |\mathbf{u}|^2, & ()' &= \frac{d}{d\tau} \\
t' &= r, & \mathbf{x} &= \mathbf{u}\,\mathbf{u}^\star & & \\
h' &= -\varepsilon\,\langle x', f(\mathbf{x}, t)\rangle & \text{or} \quad h &= r^{-1}\,(\mu - 2\,|\mathbf{u}'|^2) & &
\end{aligned}
\tag{49}
$$

Remark. Introducing the *osculating* eccentric anomaly E by $dE = \sqrt{2\,h}\,d\tau$ transforms the first differential equation into

$$4\,\mathbf{u}'' + \mathbf{u} = \frac{\varepsilon}{h}\,\left(r\,\mathbf{f}(\mathbf{x}, t)\,\bar{\mathbf{u}}^\star + 2\,\langle x', f(\mathbf{x}, t)\rangle\,\mathbf{u}'\right), \tag{50}$$

a perturbed harmonic oscillator with constant frequency $\omega = \frac{1}{2}$. As of here $()' = d/dE$. This equation is particularly well suited for introducing orbital elements with simple pertubation equations.

5.1   Osculating elements

Consider the scalar differential equation

$$4\,u'' + u = g, \quad u \in R \tag{51}$$

as a model for the system (50) of differential equations, where g stands for a small perturbation, e.g., the right-hand side of (50). With $v := 2\,u'$ Eq. (51) may be written as

$$\begin{pmatrix} u \\ v \end{pmatrix}' = A \begin{pmatrix} u \\ v \end{pmatrix} + \begin{pmatrix} 0 \\ g/2 \end{pmatrix} \quad \text{with} \quad A = \begin{pmatrix} 0 & 1/2 \\ -1/2 & 0 \end{pmatrix}. \tag{52}$$

A matrix solution $U(E)$ of the unperturbed problem $U' = A\,U$ is

$$U(E) = \exp(A\,E) = \begin{pmatrix} \cos(\dfrac{E}{2}) & \sin(\dfrac{E}{2}) \\ -\sin(\dfrac{E}{2}) & \cos(\dfrac{E}{2}) \end{pmatrix}.$$

We finally solve the perturbed system (50) by the method of *variation of the constant*. We seek a solution of the form

$$\begin{pmatrix} u(E) \\ v(E) \end{pmatrix} = U(E) \begin{pmatrix} a(E) \\ b(E) \end{pmatrix},$$

where $a(E), b(E)$ are the slowly varying orbital elements. Substituting this and its derivative into the vector differential equation (52) and solving for the derivatives of the elements yields the differential equations for the *osculating orbital elements* as functions of the eccentric anomaly E:

$$\begin{aligned} \frac{d\mathbf{a}}{dE} &= -\frac{\mathbf{g}}{2} \cdot \sin(\frac{E}{2}) \\ \frac{d\mathbf{b}}{dE} &= \frac{\mathbf{g}}{2} \cdot \cos(\frac{E}{2}) \,. \end{aligned} \tag{53}$$

Here we have used boldface symbols in order to indicate that the above differential equations for the osculating orbital elements not only hold for scalars $a, b, u, g \in R$, but also for quaternions $\mathbf{a}, \mathbf{b}, \mathbf{u}, \mathbf{g} \in U$, as well as for vectors $\vec{a}, \vec{b}, \vec{u}, \vec{g} \in R^n$, $n \in N$.

The final pertubation equations may be obtained by using $dE = \sqrt{2\,h}\,d\tau$ and (50) in order to transform the system (49) to differentiations with respect to E instead of τ. Furthermore, the equation (50) in the first line must be replaced by the system (53). In this form the equations of motion are well suited for numerical purposes. Series expansions may easily be obtained for perturbations not depending explicitly on time.

6 Conclusions

- The classical regularizations by Levi-Civita or Kustaanheimo-Stiefel transform the equations of motion of the planar or spatial Kepler problem into the linear differential equations of a harmonic oscillator in 2 or 4 dimensions, respectively.

- Based on this, we are able to formulate a simple and concise perturbation theory of the Kepler problem.

- The "language" of quaternions allows for a concise formalism for developing the Kustaanheimo-Stiefel theory of regularization of the spatial Kepler problem.

- The use of the *"star-conjugate"* $\mathbf{u}^\star$ of a quaternion $\mathbf{u}$ according to

$$\mathbf{u} = u_0 + i\,u_1 + j\,u_2 + k\,u_3\,, \quad \mathbf{u}^\star = u_0 + i\,u_1 + j\,u_2 - k\,u_3$$

yields a spatial regularization theory in perfect formal agreement with Levi-Civita's planar regularization using complex variables.

References

Brouwer, D. and Clemence, G. M., (2001), *Methods of Celestial Mechanics* (Academic Press, New York London).

Hamilton, W.R., (1844), 'On quaternions, or a new systen of imaginaries in algebra', *Philosophical Magazine* **25**, 489-495.

Hopf, H., (1931), 'Uber die Abbildung der dreidimensionalen Sphäre auf die Kugelfläche', *Mathematische Annalen* 104. Reprinted in *Selecta Heinz Hopf*, 1964, 38-63 (Springer-Verlag, Berlin Heidelberg New York).

Kustaanheimo, P., (1964), 'Spinor regularisation of the Kepler motion', *Annales Universitatis Turkuensis* A **73**, 1-7. Also *Publications of the Astronomical Observatory Helsinki* 102.

Kustaanheimo, P. and Stiefel, E. L., (1965), 'Perturbation theory of Kepler motion based on spinor regularization', *Journal für die Reine und Angewandte Mathematik* **218**, 204-219.

Levi-Civita, T., (1920), 'Sur la régularisation du problème des trois corps', *JActa Mathematica* **42**, 99-144.

Stiefel, E. L. and Scheifele, G., (1971), *Linear and Regular Celestial Mechanics* (Springer-Verlag, Berlin Heidelberg New York).

Sundman, K.F., (1907), 'Recherches sur le problème des trois corps', *Acta Societatis Scientificae Fennicae* **34**, 6..

Vivarelli, M.D., (1983), 'The KS transformation in hypercomplex form', *Celestial Mechanics and Dynamical Astronomy* **29**, 45-50.

Vrbik, J., (1994), 'Celestial mechanics via quaternions', *Canadian Journal of Physics* **72**, 141-146.

Vrbik, J., (1995), 'Perturbed Kepler problem in quaternionic form', *Journal of Physics A* **28**, 193-198.

Waldvogel, J., (2006a), 'Order and chaos in satellite encounters', in *Proceedings of the conference From Order to Disorder in Gravitational N-Body Dynamical Systems*, Cortina d'Ampezzo Italy, September 2003, editors Steves B A, Maciejewski A J, Hendry M (Springer, Dordrecht), 233-254.

Waldvogel, J., (2006b), 'Quaternions and the perturbed Kepler problem', *Celestial Mechanics and Dynamical Astronomy* **95**, 201-212.

Mechanisms for the production of chaos in dynamical systems

Massimiliano Guzzo

Dipartimento di Matematica Pura ed Applicata, Università degli Studi di Padova, Italy

1   Introduction

What is chaos? The contemporary idea of chaos in deterministic dynamical systems has its roots in the fundamental work of H. Poincaré (1892), and is related to the *real* possibility of predicting the future of a given system. While a lot of fundamental research has been done to explain the full implications of chaotic behavior, the signature of chaos has been found with the help of modern numerical integrations in many systems of the real world, and specifically of astronomy, such as the orbits of our (and others) planetary systems, of many asteroids, in the rotation of planets and satellites.

This chapter is an introduction to the basic phenomena which produce chaos in dynamical systems. First, we will explain through famous model examples, such as the celebrate Arnold cat, the Smale horseshoe, the standard map, what is usually meant by chaotic dynamics. Second, we will review some results which relate chaos to diffusion in dynamical systems, and show an application to the dynamics of the giant planets of the solar system.

A (discrete topological) dynamical system is represented by a map ψ defined on a set M:

$$\psi : M \longrightarrow M \ ,$$

such that, if $x \in M$ represents the state of the system at time $t = 0$, $\psi(x)$ represents the state of the system at the time $t = 1$. The set M of the possible states is called phase–space of the system.

A way to visualize the differences in the orbits of close initial conditions is to represent the evolution in the phase–space of a set A of initial conditions,

$$A \ , \ \psi(A) \ , \ \psi^2(A) \ , \ \dots \ , \ \psi^t(A) \ , \ \dots$$

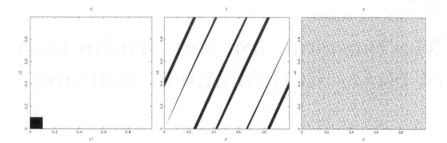

Figure 1. *Representation of $A_{0.1}$ (left panel), $\psi^3(A_{0.1})$ (center panel) and $\psi^6(A_{0.1})$ (right panel).*

rather than the evolution of single orbits. The way the sequence $\psi^t(A)$ spreads in the phase–space allows one to have a strong insight on the differences in the evolution of close initial conditions. A very instructive example is represented by the so–called 'Arnold cat' (Arnold and Avez 1967). The phase–space of this dynamical system is the torus: $M = \mathbb{T}^2$, and the orbit of a point x is given by the iteration of the map $\psi : \mathbb{T}^2 \to \mathbb{T}^2$ defined by:

$$(x_1, x_2) \longmapsto \begin{pmatrix} x_1 + x_2 \\ x_1 + 2x_2 \end{pmatrix} \mathrm{mod}(1)$$

The origin is a saddle fixed point. To visualize the orbits of points with co-ordinates near the origin, we represent the evolution of the squares $A_\rho = [0, \rho] \times [0, \rho]$, with ρ ranging from 0.1 down to the very small 10^{-14}. The result is represented in Figure 1. The evolution of $A_{0.1}$ is such that after only six iterations of the map the set is uniformly spread in the whole torus, so that, after few iterations, one is not able to distinguish the orbits of 2 points with initial conditions $|x - x'| \sim 0.1$. One could think that the 'poor control' on the orbits is due to the big distance 0.1 between initial conditions. Therefore, we now repeat the computation using a very small initial square $A_{10^{-14}}$ (the radius is smaller than atomic radii!), and represent the result in Figure 2.

In this case we observe a spread of A_ρ in only 31 iterations, even if we have decreased the value of ρ by thirteen orders of magnitude. This result shows that, no matter how small we choose a set of points, their orbits spread uniformly in the whole phase–space in a small number of iterations of the map, which actually depends logarithmically on ρ.

The ingredients which produce the strong chaotic behavior of this system (in fact among the most chaotic ones) are:

- The existence of a hyperbolic fixed point, which is a saddle point. In fact, the matrix $\frac{\partial \psi}{\partial x}(0,0)$ has real eigenvalues λ_1, λ_2 such that $0 < \lambda_1 < 1 < \lambda_2$.

- The eigenvectors u_1, u_2 have irrational slopes, so that the stable set of

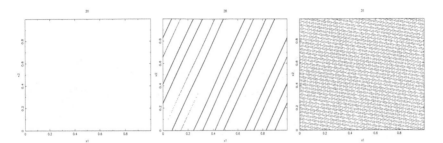

Figure 2. *Representation of $\psi^{20}(A_{10^{-14}})$ (left panel), $\psi^{28}(A_{10^{-14}})$ (center panel) and $\psi^{31}(A_{10^{-14}})$ (right panel).*

the fixed point:

$$W^s = \{(x_1, x_2) : \lim_{t \to \infty} \psi^t(x_1, x_2) = (0, 0)\} = <u_1> \mod(1)$$

and the unstable set:

$$W^u = \{(x_1, x_2) : \lim_{t \to \infty} \psi^{-t}(x) = (0, 0)\} = <u_2> \mod(1)$$

are dense, with a dense set of transverse intersections.

- The linearized map is hyperbolic at any point.

In more general cases the key to understand the chaotic behavior is represented by the structure of the stable and unstable sets. While the density of the intersection of these sets and the hyperbolicity at any point of the map are very special properties, the much more generic chaotic behavior can be reproduced when the system has a hyperbolic fixed point with stable and unstable set having an homoclinic intersection, i.e. a point of transverse intersection.

While the Arnold's cat represents a quite special example of chaotic system, the paradigm of chaos arising in conservative systems is represented by the so–called standard map $(I, \varphi) \mapsto (I', \varphi')$ defined by:

$$\begin{aligned} \varphi' &= \varphi + I \\ I' &= I + \epsilon \sin(\varphi + I) \end{aligned} \tag{1}$$

with $(I, \varphi) \in \mathbb{R} \times \mathbb{T}$ and $\epsilon \in \mathbb{R}$ is a parameter. For $\epsilon = 0$ the map is integrable: the action I is constant while the angle φ rotates with angular velocity I. For $\epsilon \neq 0$ the phase–portrait (see Figure 3) shows the presence of invariant curves, as well as two dimensional regions where motions seem to spread similarly to the case of the Arnold's cat. We remark that, also the standard map, has a hyperbolic saddle point.

Two dimensional symplectic maps, similar to the standard map, can be obtained as Poincaré sections of higher dimensional dynamical systems. An interesting example is provided by a simplified model of the spin–orbit rotations of an oblate satellite (see Figure 4).

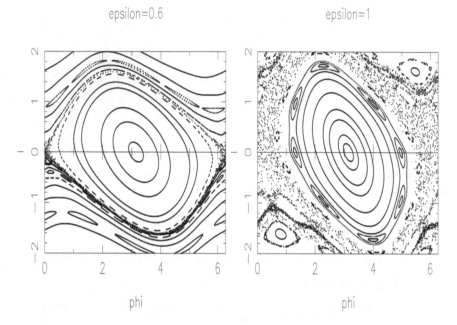

Figure 3. *Phase portrait of the standard map for $\epsilon = 0.6$ (left panel) and $\epsilon = 1$ (right panel).*

2　The homoclinic tangle of hyperbolic saddle points

In all the examples previously quoted we remarked the presence of a saddle fixed point x_*. The stable (unstable) manifold theorem (see Hirsch et al. 1977) guarantees that if a map is smooth (as in the cases of Section 1), then the sets:

$$W^s = \{x : \lim_{t \to \infty} \psi^t(x) = x_*\} \quad , \quad W^u = \{x : \lim_{t \to \infty} \psi^{-t}(x) = x_*\}$$

are smooth curves, locally tangent in x_* to the eigenvectors of $\frac{\partial \psi}{\partial x}(x_*)$. Therefore, we will call W^s the stable manifold of x_* and W^u the unstable manifold. It is evident that in two–dimensional systems with a first integral the stable and unstable manifolds are contained in the level curve of the integral containing the fixed point. Therefore, they cannot have a complicate topology. To produce a complicate structure for these manifolds we need a different hypothesis, which is the existence of a homoclinic point. In such a case, one can show that:

i) Each point of the orbit of the homoclinic point z_0:

$$z_t = \psi^t(z_0) \quad t \in \mathbb{Z}$$

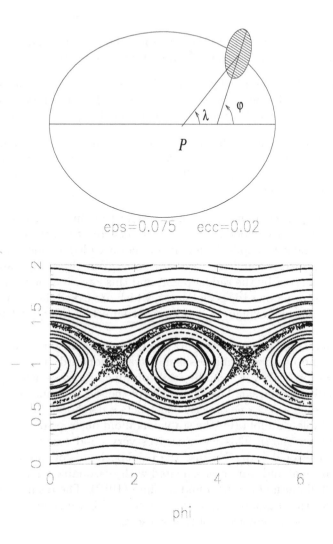

eps=0.075   ecc=0.02

phi

Figure 4. *Top: a tri-axial satellite, whose center of mass moves on a Keplerian orbit, is constrained to rotate in the plane x, y around an axis of inertia. The equation of motion for the libration angle φ is: $\ddot{\varphi} = -\frac{3}{2}\Omega^2\left(\frac{a}{|r|}\right)^3 \frac{I_2-I_1}{I_3} \sin(2(\varphi - \lambda))$, where λ, a, r denotes the true longitude, the semi–major axis and the distance from the center of mass to the central body; Ω denotes the Keplerian frequency of motion and I_1, I_2, I_3 are the principal moments of inertia. Bottom: phase portrait of the Poincaré map $(\varphi, I) = (\varphi(0), \dot{\varphi}(0)) \longmapsto (\varphi', I') = (\varphi(T), \dot{\varphi}(T))$, with $T = 2\pi/\Omega$. The eccentricity of the orbit is 0.02 and $\frac{3}{2}\Omega^2 \frac{I_2-I_1}{I_3} = 0.075$.*

is homoclinic, i.e. $z_t \in W^s \cap W^u$ and the intersection is transverse.

ii) As a consequence of *(i)*, the unstable manifold cuts the stable manifold transversely infinite times forming typical lobes (see Figure 5).

iii) Approaching the fixed point the base of each lobe becomes smaller and the height becomes bigger (near the fixed point there is contraction along the stable direction and expansion along the unstable one).

iv) At a given point, the lobes of the unstable manifold are so big that they are forced to intersect the stable manifold in points z', z'' which are not in the orbit of z_0. The orbits of these points contain only homoclinic points (see Figure 6).

All these properties allow one to understand that in the hypothesis of existence of at least a homoclinic point the structure of the stable and unstable manifolds is indeed complicate, and is commonly called homoclinic tangle. The idea is to use this complicate structure to explain the complicate dynamics which is generated by the saddle point, and this can be done using the conjugation of the dynamics to special maps, which are called 'horseshoe' maps. In the next section we explain in detail the dynamics of the most famous Smale horseshoe map.

In general, the analytic computation of approximations of finite pieces of the stable and unstable manifolds is not straightforward. Instead, the numerical localization of W^s, W^u can be easily obtained by numerically propagating initial conditions in a small neighborhood of the saddle point up to some finite time T. In such a way, one directly constructs a neighborhood of a finite piece of the unstable manifold (for the stable manifold one repeats the construction for the inverse map). This method gives very good results for fixed points of two dimensional maps, because the neighborhoods of the fixed points are two dimensional and can be propagated with reasonable CPU times. A more sophisticated method can be found in Simo (1989). The result of such a computation for the standard map is represented in Figure 7, which gives the idea of the topological complexity of the two sets.

3   The Smale horseshoe

The famous Smale horseshoe (Smale 1967) allows one to describe the peculiar characteristics of chaotic dynamics due to hyperbolic behavior, and it is representative of the dynamics in the homoclinic tangles of hyperbolic saddle points.

The map Φ defining the Smale horseshoe is defined in the set $M = D_1 \cup R \cup D_2 \subseteq \mathbb{R}^2$, where R is a rectangle of base α and height β and D_1, D_2 are half-disks like the ones represented in Figure 8. Let Φ be defined by $\Phi = \phi_2 \circ \phi_2 : M \to M$, where ϕ_1 expands the base of R by a factor $a > 2$ and

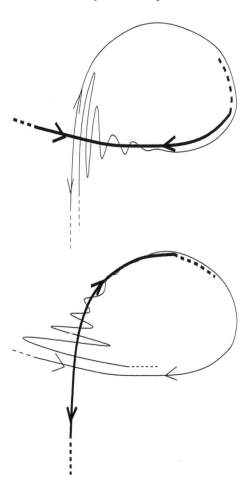

Figure 5. *Top: the unstable manifold cuts the stable manifold transversely infinite times forming typical lobes. Bottom: the stable manifold cuts the unstable manifold transversely infinite times forming typical lobes.*

contracts its height by a factor $b < 1/a < 1/2$, and ϕ_2 bends $\phi_1(R)$ like a horseshoe (see Figure 8).

Let us remark that the map Φ has a fixed point in D_1, which attracts any orbit with initial condition in D_1, and therefore in D_2 and any initial condition $x_0 \in R$ such that for some $k \in \mathbb{N}$ it is $\Phi^k(x_0) \in D_1 \sqcup D_2$. Therefore, the most interesting dynamics occurs for those points $x_0 \in R$ such that $\Phi^k(x_0) \in R$ for any $k \in \mathbb{N}$, that is for all points in the set:

$$\Lambda_+ = \cap_{k \geq 0} \Phi^{-k}(R) \ , \quad \phi^{-k}(R) = \{x \in R : \Phi^k(x) \in R\} \ .$$

Because we want to find a set on which the map Φ defines a dynamical system,

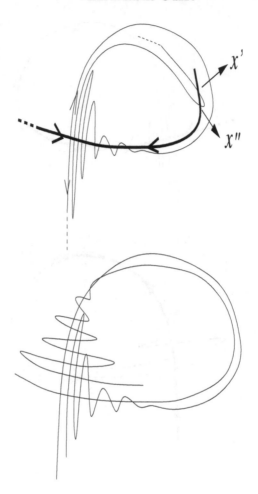

Figure 6. *Top: homoclinic points which are not in the orbit of x_0. Bottom: homoclinic tangle near the fixed point.*

we restrict Λ_+ to the set $\Lambda = \Lambda_+ \cap \Lambda_-$ where:

$$\Lambda_- = \cap_{k \geq 0} \Phi^k(R) \ .$$

The set Λ has a peculiar geometry. To understand it, we first remark that the set:

$$\Lambda_1^+ = \{x \in R : \quad \Phi(x) \in R\} = \Delta_0 \cup \Delta_1$$

is made by two vertical rectangles Δ_0, Δ_1 of base α/a as in Figure 9, left panel; the set:

$$\Lambda_2^+ = \{x \in \Lambda_1^+ : \quad \Phi^2(x) \in R\} = \cup_{i,j \in \{0,1\}} \Delta_{i,j}$$

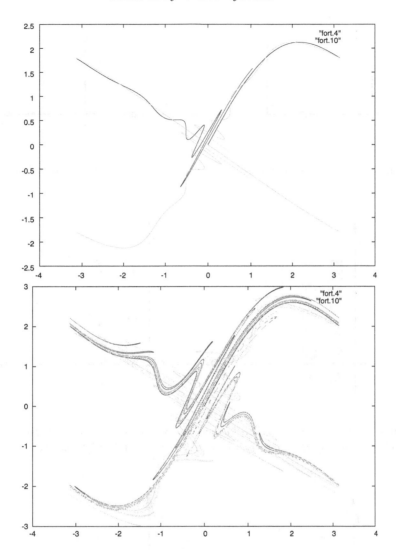

Figure 7. *Numerical representation of finite pieces of the stable and unstable manifolds of the hyperbolic fixed point of the standard map. The perturbing parameter is $\epsilon = 1$. The upper figure represents a shorter piece of the manifold. The horizontal axis corresponds to the angle $\varphi \in [-\pi, \pi]$, while the vertical axis corresponds to the action I.*

is made by the four vertical rectangles $\Delta_{i,j}$ of base α/a^2 (see Figure 9, right panel); Λ_k^+ is made by 2^k vertical rectangles of base α/a^k. Therefore, the set $\Lambda_+ = \cap\Lambda_k^+$, is made by an infinite number of vertical lines arranged as a Cantor set. The geometry of the set Λ^- is very similar to the geometry of Λ_+: the set $\Lambda_1^- = R \cap \Phi(R) = \tilde{\Delta}_0 \cup \tilde{\Delta}_1$ is made by the horizontal rectangles

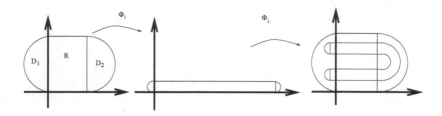

Figure 8. *The maps ϕ_1 and ϕ_2 map the set $M = D_1 \cup R \cup D_2$ to a horseshoe.*

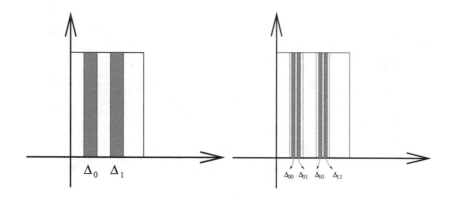

Figure 9. *The set Λ_1^+ is made by the two rectangles Δ_0, Δ_1 (left panel). The set Λ_2^+ is made by the four rectangles $\Delta_{i,j}$, $i, j = 0, 1$ (right panel).*

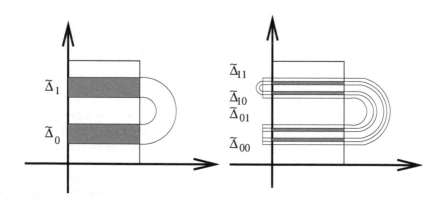

Figure 10. *The set Λ_1^- is made by the two rectangles $\tilde{\Delta}_0, \tilde{\Delta}_1$ (left panel). The set Λ_2^- is made by the four rectangles $\tilde{\Delta}_{i,j}$, $i, j = 0, 1$ (right panel).*

$\tilde{\Delta}_0, \tilde{\Delta}_1$ (see Figure 10, left panel); the set:

$$\Lambda_2^- = R \cap \Phi(R) \cap \Phi^2(R) = \cup_{i,j \in \{0,1\}} \tilde{\Delta}_{ij}$$

is made by the horizontal rectangles $\tilde{\Delta}_{i,j}$ (see Figure 10, right panel); Λ_k^- is made 2^k by the horizontal rectangles; the set $\Lambda_- = \cap \Lambda_k^-$ has the structure of a Cantor set of horizontal lines.

As a consequence, the set $\Lambda = \Lambda^+ \cap \Lambda^-$ has the structure of a double Cantor set: each point $x \in \Lambda$ is in the intersection of a vertical line of Λ_+ and an horizontal line of Λ_-.

The properties of the dynamics in Λ are usually explained by means of a symbolic representation. Precisely, one considers the set:

$$\Sigma = \{\sigma = (....\sigma_{-1}, \sigma_0, \sigma_1, ...) \ , \sigma_k \in \{0,1\} \ \forall k\}$$

of the double infinite sequences of symbols $0, 1$, and define a bijective map $h : \Lambda \longrightarrow \Sigma$. The conjugate map $\tilde{\Phi} : \Sigma \to \Sigma$, $\tilde{\Phi} = h \circ \Phi \circ h^{-1}$ is called symbolic dynamics of the map Φ. Among all possible symbolic representations of Φ, one of the most useful is defined as follows: for any $z \in \Lambda$, the sequence $h(z)$ is defined by

$$\Phi^t(z) \in \Delta_{h(z)_t}$$

for any $t \in \mathbb{Z}$. In other words, the sequence $h(z)$ is constructed by declaring explicitly in which rectangle Δ_0, Δ_1 are all the points in the orbit of z. With this symbolic conjugation, the symbolic dynamics $\tilde{\Phi}$ is a very simple map defined by:

$$\tilde{\Phi}(\sigma)_k = \sigma_{k+1} \ ,$$

i.e. it is a left translation of the symbols of σ. The dynamics of the left translation can be easily studied. For example, the map has the fixed points $\sigma, \eta \in \Sigma$ defined by:

$$\sigma_k = 0 \ , \ \forall \ k \ , \ \eta_k = 1 \ , \ \forall \ k \ .$$

Moreover, any periodic sequence σ of period T is the initial condition of a periodic orbit of the same period. Correspondingly, one obtains fixed points and periodic orbits of the Smale horseshoe map. Also, one can construct orbits of the Smale horseshoe map such that $\Phi^k(z)$ are in Δ_{σ_k} for any sequence σ_k. Defining a suitable distance on Σ, such as the distance:

$$\delta(\sigma', \sigma'') = 2^{-K}$$

where $K \in \mathbb{N}$ is such that $\sigma'_j = \sigma''_j$ for any $|j| \leq K - 1$, and $\sigma'_K \neq \sigma''_K$ or $\sigma'_{-K} \neq \sigma''_{-K}$ (if $\sigma' - \sigma''$ it is $\delta(\sigma', \sigma'') = 0$), then h is a continuous homeomorphism. With reference to the distance δ, one shows that the periodic orbits of $\tilde{\Phi}$ are dense in Σ, and that there exist points whose orbit is dense in the phase space. Because h is an homeomorphism, the same properties hold for the Smale horseshoe map. All these facts characterize the complexity of the dynamics of the Smale horseshoe map, and generically we say that it is a chaotic dynamics.

The set Λ is hyperbolic. Hyperbolicity is strictly related to all the peculiar properties of this map, and it is often used to define chaos. The measure of hyperbolicity of a given dynamical system $\psi : M \to M$ is represented by the so–called Lyapunov characteristic exponents. The Lyapunov characteristic exponent related to the initial condition $x_0 \in M$ and to the initial tangent vector $\xi_0 \in T_{x_0} M$ is defined by the limit:

$$\chi(x_0, \xi_0) = \lim_{t \to \infty} \frac{1}{t} \ln \|\xi_t\| \ , \tag{2}$$

where $\xi_t = \frac{\partial \psi^t}{\partial x}(x_0)\xi_0$. The largest Lyapunov characteristic exponent, i.e.

$$\chi(x_0) = \sup_{\xi \neq 0} \chi(x_0, \xi_0) \ ,$$

related to the maximum hyperbolic local expansion around x_0, is frequently used to characterize chaos. Precisely, an orbit is considered to be chaotic if its largest Lyapunov exponent is strictly positive.

4   Chaotic dynamics in the homoclinic tangles

In this section we show how to transport the results obtained for the Smale horseshoe map (or similar maps) to homoclinic tangles related to saddle fixed points. Let us consider the dynamical system defined by the map $\psi : M \to M$, with $M \subseteq \mathbb{R}^2$, having a saddle fixed point z and a homoclinic point z_0 as in Figure 11, top–left panel. Let us define a set R which plays the role of the rectangle R of the horseshoe map (see Figure 11 top–right panel): specifically, in a given number t of iterations the set $\psi^t(R)$ intersects R as it is shown in Figure 11, bottom–right panel. The map ψ^t acts on R as a horseshoe map, and apart from some topological details (for example the sets Δ_0, Δ_1, $\tilde{\Delta}_0$, $\tilde{\Delta}_1$ are not exactly rectangles as in the original horseshoe map) the same conclusions apply. In particular, it is possible to define the conjugation to a symbolic dynamics represented by the left–shift of double infinite sequences. In conclusion:

- non-trivial, chaotic dynamics are produced near the homoclinic tangles of saddle fixed points;

- the existence of a homoclinic point is a minimum requirement for the existence of a complicate structure for the stable and unstable manifolds of the fixed point;

- it is possible to identify horseshoes (or similar maps) for some iterates of the dynamical systems, so that one can describe the dynamics by means of the conjugation to a symbolic dynamics;

- symbolic dynamics, such as the left–shift map, are characterized by the existence of dense sets of periodic orbits, of dense orbits, of orbits which arbitrarily occupy specific sets of the phase space.

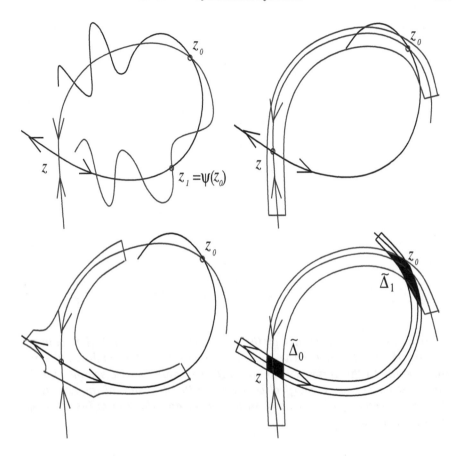

Figure 11. *Top–left panel: stable and unstable manifolds of a saddle fixed point z with a homoclinic point z_0. Top–right panel: definition of the set R which plays the role of the rectangle R of the horseshoe map. Bottom–left panel: the map ψ contracts R along the stable manifold and expands it along the unstable manifold. Bottom–right panel: in a given number t of iterations the set $\psi^t(R)$ intersects R in two sets $\tilde{\Delta}_0$, $\tilde{\Delta}_1$.*

5 From chaos to diffusion in two dimensional systems

The existence of chaos in a dynamical system does not mean that the system is characterized by macroscopic instability. In fact, the region interested by chaotic motions can be well localized in the phase–space. For example, we consider the standard–like map:

$$\varphi' \;=\; \varphi + I$$

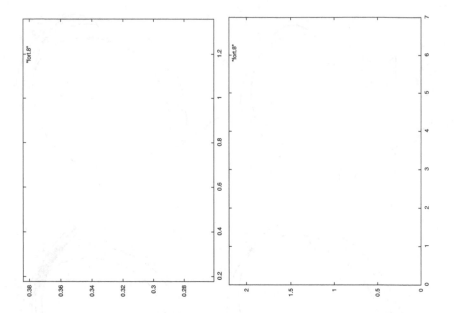

Figure 12. *Phase portraits of the map (3) for ε = 0.002 (left panel) and ε = 0.004 (right panel). The x axis corresponds to the angle φ, the y axis corresponds to the action I. For the smallest value of ε we see that the KAM invariant curves are topological barriers for the diffusion of the action I. For the largest value of ε we see that in principle, an orbit can diffuse through all the action values.*

$$I' \;=\; I + \epsilon \frac{\sin \varphi'}{(\cos \varphi' + 1.1)^2} \; . \tag{3}$$

For $\epsilon = 0.002$ there are many regions of chaotic motions in the phase space (see Figure 12, left panel). However, all these regions are disconnected and bounded by invariant curves. Because invariant curves are complete barriers to diffusion, despite the presence of many chaotic regions the action variable I cannot diffuse through a large interval of values. The invariant curves which are complete barriers to the diffusion of the action variable are KAM curves, whose existence is established by the KAM theorem (Kolmogorov 1954, Arnold 1963, Moser 1958) for small values of the perturbing parameter. Instead, for higher values of the perturbing parameter these invariant curves typically are destroyed and replaced by the so called cantori, which are discontinuous invariant sets, and therefore the action variables can diffuse through their holes. In Figure 12, right panel, we see that the situation is very different from Figure 12, left panel, because the chaotic zone allows, in principle, an orbit to diffuse through all the action values. However, the model for chaos based on the homoclinic tangles is not sufficient to explain the existence of such a global diffusion. To generalize it we need to define the heteroclinic tangles related to hyperbolic periodic orbits.

A periodic orbit of period k is defined by the points $x_j \in M$, $j = 0, .., k-1$, such that $\psi(x_j + k) = \psi(x_{j+k})$ for any j. It is evident that the periodic orbits of period k can be found as the fixed points of the map ψ^k, and we consider the periodic orbit of initial conditions x_0 such that x_0 is a saddle fixed point of ψ^k. One can therefore define stable and unstable manifolds as in the case of fixed points, and look for homoclinic intersections, producing chaotic motions. But now we can do more. Let us consider for simplicity the standard map ψ, and its periodic orbits of period 2. We already know that $(0, 0)$ is a saddle fixed point of ψ, but we find also a saddle fixed point $x_0 = (\varphi_0, I_0)$ of ψ^2 which corresponds to a hyperbolic periodic orbit of the standard map of period 2 (see Figure 13, left panel). For $\epsilon = 2$, there are no invariant KAM curves which are complete barriers to the diffusion of the action variable, but one needs a mechanism for this diffusion to effectively take place. This mechanism is provided by the transverse intersections among the stable manifold of one periodic orbit and the unstable manifold of the other one. Such transverse intersection points are called 'heteroclinic points,' and their existence is used since decades to explain the presence of global diffusion in the phase space. In Figure 13, right panel, we report the numerical computation of the unstable manifold of $(0, 0)$ and the stable manifold of x_0: the existence of transverse intersections is evident, and therefore one can select initial conditions which chaotically diffuse from a neighborhood of $(0, 0)$ to a neighborhood of x_0. This mechanism of diffusion could be considered as the topological mechanism explaining the celebrate Chirikov diffusion (Chirikov 1979), which occurs when two resonant regions overlap.

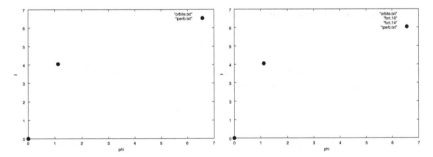

Figure 13. *On the left: phase portrait of the standard map for $\epsilon = 2$. The bullet denotes the initial condition of a hyperbolic periodic orbit of period 2. On the right: on the phase portrait we report the computation of a finite piece of the unstable manifold of $(0, 0)$ and the stable manifold of the hyperbolic periodic orbit. Many heteroclinic intersections are evident.*

6   Diffusion in higher dimensional systems: the Arnold's model for diffusion

If the mechanisms for chaos and diffusion in two dimensional maps are well understood, a lot of research is still currently done to characterize the mechanisms for diffusion in the higher dimensional systems. In fact, apart from specific examples, the understanding of the general mechanisms which can produce drift and diffusion in the phase space of such systems is an interesting, and in general, open problem. In this section we describe the pioneering example proposed by Arnold (1964), which inspired many works through the last decades.

The Arnold's example is defined by the following Hamiltonian system:

$$H = \frac{I_1^2}{2} + \frac{I_2^2}{2} + \epsilon \cos \varphi_1 + \epsilon \mu (\cos \varphi_1 - 1)(\sin \varphi_2 + \sin t) \tag{4}$$

whose Hamilton equations are:

$$\begin{aligned}
\dot{\varphi}_1 &= I_1 \\
\dot{\varphi}_2 &= I_2 \\
\dot{I}_1 &= \epsilon \sin \varphi_1 + \epsilon \mu \sin \varphi_1 (\sin \varphi_2 + \sin t) \\
\dot{I}_2 &= -\epsilon \mu (\cos \varphi_1 - 1) \cos \varphi_2 \ .
\end{aligned} \tag{5}$$

The system depends on two parameters ϵ and μ. For $\epsilon = 0$ the system has only 3D invariant tori (we consider t as a periodic variable with equation $\dot{t} = 1$):

$$\dot{\varphi}_1 = I_1 = \ const \qquad \dot{\varphi}_2 = I_2 = \ const \qquad \dot{t} = 1 \ .$$

For $\epsilon \neq 0$ we consider the special resonance:

$$\dot{\varphi}_1 = 0$$

which contains an invariant manifold Λ defined by the equations $I_1 = 0, \varphi_1 = 0$, which is fibered by the invariant 2D tori:

$$I_1 = 0 \qquad \varphi_1 = 0 \qquad \dot{\varphi}_2 = I_2(0) \qquad \dot{t} = 1 \ ,$$

which are hyperbolic. In fact, for $\mu = 0$ the system is the product of a pendulum and a rotator:

$$H = \frac{I_1^2}{2} + \epsilon \cos \varphi_1 + \frac{I_2^2}{2} \ .$$

The stable and unstable manifolds of each torus are the separatrices of the pendulum. As a consequence, for $\mu = 0$, the invariant manifold Λ is hyperbolic but we remark that there is no diffusion of the action I_2 on Λ. From general hyperbolic theory, the invariant manifold remains hyperbolic for sufficiently small values of μ. In such a case, Arnold proved that the unstable manifolds of hyperbolic tori of Λ intersect transversely the stable manifolds of close hyperbolic tori of Λ, thus providing the mechanism for initial conditions in the neighborhood of Λ to diffuse through these heteroclinic points. This kind of diffusion is called 'Arnold diffusion.'

Though the results on the Arnold's model have not been generalized to generic quasi–integrable Hamiltonian systems, the ideas contained in Arnold's work are used to study the problem of diffusion in higher dimensional systems. Let us quickly review these ideas. We consider a dynamical system defined by a symplectic map $\psi : M \to M$, M being a symplectic and Riemannian manifold, which has a hyperbolic invariant manifold $\Lambda \subseteq M$ (precisely, Λ is normally hyperbolic, see Hirsch et al. 1977) and the dynamics of ψ restricted to Λ has no diffusion, for example any orbit of $\phi_{|\Lambda}$ is uniformly bounded by some constant c_0. The diffusion concerns therefore a neighborhood of Λ. Specifically, one would like to find for any $c > c_0$ and $\rho > 0$, two points x, y up to a distance ρ from Λ which are on the same orbit and their distance is greater than c. A way of proving the existence of this kind of diffusion is to prove that the stable and unstable manifolds of the invariant objects of Λ intersect transversely. In quasi–integrable systems this kind of diffusion is usually called Arnold diffusion. In the next section we provide an example of Arnold diffusion in a quasi–integrable symplectic 4D map, which is more generic than Arnold's model.

7 Arnold diffusion in a quasi integrable 4D system

In many papers in collaboration with Froeschlé and Lega (2000) we considered a specific quasi–integrable map as a model system to study diffusion in 4D systems, which is more generic than Arnold's model. The map consists of two coupled twist maps as follows:

$$\varphi_1' = \varphi_1 + I_1 \quad , \quad \varphi_2' = \varphi_2 + I_2$$
$$I_1' = I_1 - \epsilon \frac{\partial f}{\partial \varphi_1}(\varphi_1', \varphi_2') \quad , \quad I_2' = I_2 - \epsilon \frac{\partial f}{\partial \varphi_2}(\varphi_1', \varphi_2') \tag{6}$$

where ϵ is 'small' and:

$$f = \frac{1}{\cos(\varphi_1) + \cos(\varphi_2) + c}$$

with $c > 2$. If $\epsilon = 0$ the actions of this system are constants of motion and the angles rotate at constant angular velocity. It will be convenient to represent the dynamics of the map on the surface:

$$S = \{(I_1, I_2, \varphi_1, \varphi_2) : \ (\varphi_1, \varphi_2) = (0, 0)\}.$$

For $\epsilon = 0$, any initial condition (I_1, I_2) on S does not return on S, or it returns exactly on (I_1, I_2). Therefore, each orbit with initial condition on S

is symbolically represented by a dot on S. If $\epsilon \neq 0$ one does not expect that the actions are constants of motion, but if ϵ is sufficiently small, the phase space is filled by a large volume of 2D KAM tori. In particular, anyone of these tori intersects transversely S only on one point (see Guzzo et al 2002), and therefore each invariant torus is symbolically represented by a point on S. Therefore, the section S contains possibly many points belonging to invariant tori, but these 2D invariant tori do not trap motions in the 4D phase space: there is the possibility of diffusion even for very small ϵ (see Figure 14 for a symbolic representation of possible diffusion paths).

Diffusion, as far as we know, needs hyperbolic structures, which are related to the resonances of the system, therefore we need a method to identify the hyperbolic structures in the resonances. We first recall the definition of resonances for this system. The angle $k_1 \varphi_1 + k_2 \varphi_2$, with $k_1, k_2 \in \mathbb{Z}$, is resonant when it exists $k_3 \in \mathbb{Z}$ such that:

$$ k_1 \varphi_1' + k_2 \varphi_2' = (k_1 \varphi_1 + k_2 \varphi_2) + (k_1 I_1 + k_2 I_2) = (k_1 \varphi_1 + k_2 \varphi_2) + 2\pi k_3 \ , $$

i.e. when:

$$ k_1 I_1 + k_2 I_2 - 2\pi k_3 = 0 \ . $$

The KAM theorem tells us that invariant tori are located far from a suitable neighborhood of all these resonances (see Figure 14). In fact, a KAM torus exists near the actions (I_1, I_2) satisfying a suitable Diophantine condition of the form:

$$ |k_1 I_1 + k_2 I_2 - 2\pi k_3| \geq \frac{\mathcal{O}(\sqrt{\epsilon})}{|(k_1, k_2, k_3)|^\tau} \ , \quad \forall (k_1, k_2, k_3) \neq (0, 0, 0) \ . \quad (7) $$

The complement of the set of invariant tori, which is in the neighborhood of the resonances, is called 'Arnold web,' and contains the hyperbolic structures which possibly support chaotic motions and diffusion. An efficient way of detecting numerically the Arnold web of a system is provided by the so called Fast Lyapunov Indicator (see Froeschlé et al 1997; Froeschlé et al 2000). In Froeschlé et al (2000) we used this indicator to locate precisely the KAM tori and the resonances on S. Precisely, for any point of a grid of initial conditions of S we computed the Fast Lyapunov Indicator, and represented it with a color scale. For a definition and a review of the properties of the FLI we refer to Guzzo et al (2002). Here, we just remark that the result of the numerical computation of the FLI has the following interpretation:

- the points with the higher values of the FLI (which corresponds to white in the color scale used to represent the value of the indicator) are chaotic resonant motions;

- the points with an intermediate value of the FLI (which corresponds to intermediate gray in the color scale used to represent the value of the indicator) are regular motions (including KAM tori);

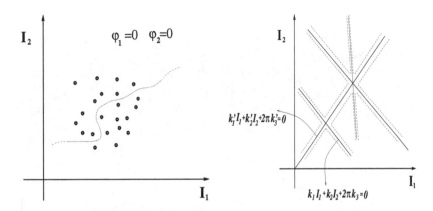

Figure 14. *On the left: the KAM tori do not trap motions in the 4D space. On the right: the KAM tori are outside the neighborhood of the resonances defined by the Diophantine condition (7).*

- the points with lower value of the FLI (which corresponds to black or dark gray in the color scale used to represent the value of the indicator) are regular motions (including resonant tori).

Therefore, the color representation of the FLI on S allows one to clearly identify the KAM tori, the resonant tori, as well as the hyperbolic motions which possibly diffuse in the phase space. The result of the computation is reported in Figure 15. For $\epsilon = 0.1$ (top left panel) there is a prevalence of KAM tori in the phase space, and the hyperbolic structures are organized as a web of resonances, as predicted by the KAM theorem. For $\epsilon = 0.6$ (top right panel) there is still a prevalence of KAM tori in the phase space, but the resonant regions are enlarged with respect to the previous case. For $\epsilon = 1.6$ there is a prevalence of resonant motions, resonant regular and chaotic. The hyperbolic structures are not organized in a web, therefore we expect that we are not in a regime described by the KAM theorem. Finally, we show in the bottom right panel a closer look at the hyperbolic manifolds around the crossing of resonances. We can clearly see that these manifolds are characterized by peculiar flower like structures, which are the analog in the higher dimensional case of the lobes of stable unstable manifolds of saddle points (Guzzo et al. 2007).

We now show how diffusion takes place on these hyperbolic manifolds. In Guzzo et al. (2005) we have shown that indeed, initial conditions in the hyperbolic manifolds diffuse in the Arnold web.

We have chosen initial conditions in the region of resonant chaotic motions far from the crossings of resonances, which we identified with the FLI method. Precisely, we have chosen twenty initial conditions near $(I_1, I_2) = (1.71, 0.81)$ and then we computed numerically their orbits the map up to 10^{11} iterations.

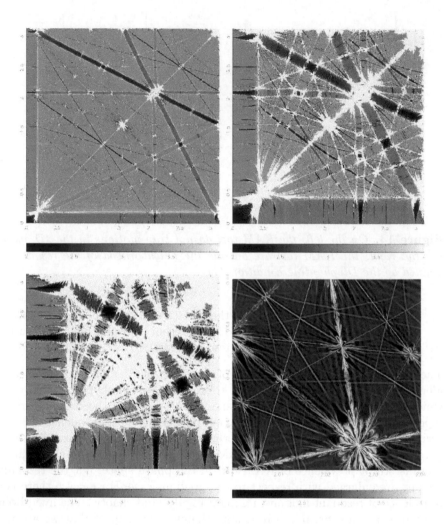

Figure 15. *Computation of the Arnold web on the section S using a color representation of the FLI. On the top left: $\epsilon = 0.1$, $c = 4$. On the top right: $\epsilon = 0.6$, $c = 4$. On the bottom left: $\epsilon = 1.6$, $c = 4$. On the bottom right: a closer look to the hyperbolic manifolds for $\epsilon = 1.d - 4$, $c = 2.1$.*

The results are reported in Figure 16: on the action plane (I_1, I_2) we plotted as black dots all points of the orbits which have returned after some time on the section S. Of course, because computed orbits are discrete we represented the points which enter the neighborhood of S defined by $|\varphi_1|, |\varphi_2| \leq 0.005$, (reducing the tolerance 0.005 reduces only the number of points on the section, but does not change their diffusion properties). In such a way we represent the diffusion properties for the initial conditions in a neighborhood of S.

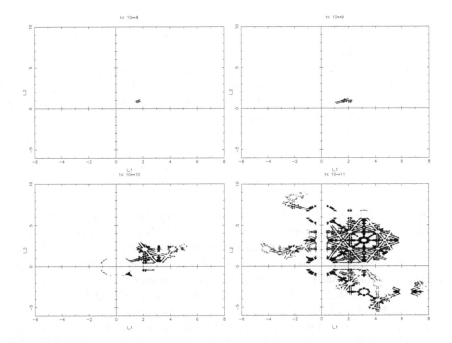

Figure 16. *Evolution on section S (black dots) of 20 orbits for the mapping (2) on a time $t < 10^8$ iterations (top left), $t < 10^9$ iterations (top right), $t < 10^{10}$ iterations (bottom left), $t < 10^{11}$ iterations (bottom right) for $\epsilon = 0.6$.*

It is evident that the orbits fill a macroscopic region of the action plane whose structure is that of the Arnold web. The possibility of visiting all possible resonances is necessarily limited by finite computational times.

8 An application to our planetary system

In the last decades the problem of the stability of our solar system has been studied by means of long–term numerical integrations (Sussman and Wisdom 1988; Laskar 1989; Nobili et al. 1989; Sussman and Wisdom 1992; Laskar 1996), which demonstrated that while the system does not have significant instabilities (especially for the outer planets), it is nevertheless chaotic. Laskar (1990) found secular resonances responsible for the chaos of the inner planets. The chaos in the orbit of the outer planets is instead due to mean–motion resonances among three or four planets Murray and Holman (1999), Guzzo (2005), Guzzo (2006), which occur when there exist at least three integers n_i, n_j, n_k such that:

$$n_i \dot{\lambda}_i + n_j \dot{\lambda}_j + n_k \dot{\lambda}_k \sim 0 \ , \tag{8}$$

more precisely when this quantity is of the same order of the secular frequencies (λ_j denotes the mean longitude of the j-th planet). As it is shown in Guzzo (2005), (2006), these resonances are arranged as a web in the space of the semi–major axes, as in the Arnold web shown in Section 6.

To better understand the problem we first show the detection of the 5–2 mean motion resonance among Jupiter and Saturn in the simplified problem represented by the three body problem Sun–Jupiter–Saturn, on a two dimensional grid of initial conditions, obtained by keeping constant all initial values of the orbital elements of the two planets, except for the initial values of their semi–major axes. The choice is motivated by the fact that semi–major axes are the elements which mainly determine the location of the mean motion resonances.

As is well known since the nineteenth century, the motion of Saturn and Jupiter is affected by the quasi–resonance 5-2, which is defined by $2\dot{\lambda}_5 - 5\dot{\lambda}_6 \sim 0$. Actually, the Sun–Jupiter–Saturn system is not in this resonance, because on average the critical angle $2\lambda_5 - 5\lambda_6$ advances monotonically with respect to time. The fact that the system is near this resonance, but not in it, can be seen in Figure 17, left panel. The 5-2 resonance is clearly identified as the large white band of chaotic motions, while the initial condition corresponding to the 'true' solar system is definitely outside this chaotic band (characterized in the picture by the gray value of the FLI). The numerical integration has been done using a corrected four order symplectic integrator described in Guzzo (2001). The step size of 0.1 yr does not allow one to represent correctly all details of planetary dynamics, but it is sufficient to properly locate the three–planet resonances, as it is explained in Guzzo (2005) and (2006).

We now also include in the system Uranus and Neptune. These two planets introduce new resonances, which can be seen in Figure 17, right panel. Beside the most important 5–2 resonance, many other resonances constitute a web in the semi–major axis space, and all these new resonances are related to the orbital elements of Jupiter, Saturn, and at least one planet among Uranus and Neptune.

In Figure 18 we represent the computation of the three and four planet resonances in different sections of the phase space. On the left the section S is obtained by fixing the initial conditions of all the orbital elements, and by changing only the initial value of the semi–major axis of Saturn and Uranus, while the right panel is obtained by changing only the semi–major axis of Uranus and Neptune. While the picture on the left is similar to the pictures representing the Arnold web of quasi–integrable systems (such as those represented in Figure 15), the picture on the right shows resonance overlapping producing a region of chaotic motions which extends for about 10^{-3} AU, and therefore the evolution of fictitious planetary systems with initial conditions in this chaotic region can undergo, in principle, very slow unstable variations of the semi–major axes of Uranus and Neptune of 10^{-3} AU.

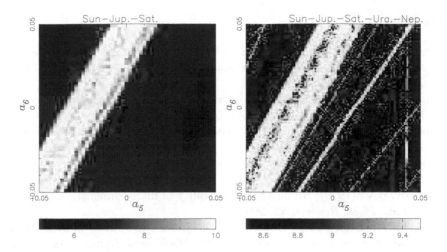

Figure 17. *Representation of mean–motion resonances using a section which is obtained by fixing the initial conditions of all the orbital elements, and by changing only the initial value of the semi–major axis of Jupiter a_5 and Saturn a_6. The coordinates of the pictures represent the differences with respect to the values of the true solar system at a given epoch, which is represented by a black bullet in $(0,0)$. The picture on the left is obtained by integrating the three body problem Sun–Jupiter–Saturn, while the picture on the right is obtained by integrating the three body problem Sun–Jupiter–Saturn–Saturn–Uranus.*

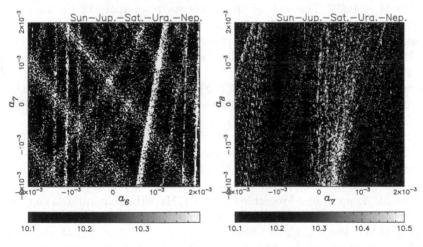

Figure 18. *The web of three planet resonances computed in the space of the semi–major axis of Saturn and Uranus (left panel) and Uranus and Neptune (right panel). The coordinates of the pictures represent the differences of the semi–major axis of the planets with respect to their 'true' values at a given epoch.*

References

Arnold V.I. (1963), Proof of a theorem by A.N. Kolmogorov on the invariance of quasi-periodic motions under small perturbations of the Hamiltonian. *Russ. Math. Surv.*, **18**, 9.

Arnold V.I. (1964), Instability of dynamical systems with several degrees of freedom, *Sov. Math. Dokl.*, **6**, 581–585.

Arnold V.I., A. Avez (1967), Problèmes Ergodiques de la Mécanique Classique. Paris, Gauthier-Villars.

Arnold V.I. (1976) Méthodes Mathématiques de la Méchanique Classique. MIR, Moscow.

Chirikov, B.V. (1979), An universal instability of many dimensional oscillator system. *Phys. Reports*, **52**, 265.

Froeschlé C., Lega E. and Gonczi R. (1997), Fast Lyapunov indicators. Application to asteroidal motion. *Celest. Mech. and Dynam. Astron.*, **67**, 41–62.

Froeschlé C., Guzzo M. and Lega E. (2000), Graphical evolution of the Arnold web: From order to chaos, *Science*, **289**, 5487.

Froeschlé C., Guzzo M. and Lega E. (2005), Local and global diffusion along resonant lines in discrete quasi–integrable dynamical systems, *Celest. Mech. and Dynam. Astron.*, **92**, 1-3, 243-255.

Froeschlé C., Lega E. (2000), On the structure of symplectic mappings. The Fast Lyapunov Indicator: a very sensitive tool, *Celest. Mech. and Dynam. Astron.*, **78**, Issue 1/4, 167-195.

Guzzo M. (2005), The web of three–planets resonances in the outer solar system, *Icarus*, **174**, n. 1., 273-284.

Guzzo M. (2006), The web of three-planet resonances in the outer solar system II: A source of orbital instability for Uranus and Neptune, *Icarus*, **181**, 475-485.

Guzzo M., Lega E. and Froeschlé C. (2002), On the numerical detection of the effective stability of chaotic motions in quasi-integrable systems, *Physica D*, **163**, Issues 1-2, 1-25.

Guzzo M., Lega E. and Froeschlé C. (2005), First numerical evidence of Arnold diffusion in quasi–integrable systems, *DCDS B*, **5**, 3.

Guzzo M., Lega E. and Froeschlé C. (2006), Diffusion and stability in perturbed non-convex integrable systems. *Nonlinearity*, **19**, 1049–1067.

Guzzo, M., Lega E. and Froeschlé C. (2007), Hyperbolic manifolds supporting Arnold diffusion in quasi-integrable systems. Preprint.

Hirsch M.W., Pugh C.C. and Shub M. (1977), Invariant Manifolds. Lecture Notes in Mathematics, **583**. Springer-Verlag, Berlin-New York.

Kolmogorov, A.N. (1954), On the conservation of conditionally periodic motions under small perturbation of the Hamiltonian, *Dokl. Akad. Nauk. SSSR*, **98**, 524.

Laskar, J. (1989), A numerical experiment on the chaotic behaviour of the solar system, *Nature*, **338**, 237–238.

Laskar, J. (1990), The chaotic motion of the solar system - A numerical estimate of the size of the chaotic zones. *Icarus* 88, 266–291.

Laskar, J. (1996), Marginal stability and chaos in the solar system (Lecture). In: Ferraz–Mello, S., Morando, B., Arlod, J. (Eds.), Dynamics, Ephemerides, and Astrometry of the Solar System: proceedings of the 172nd Symposium of the International astronomical Union, held in Paris, France, 38 July, 1995, 75–88.

Lega E., Guzzo M. and Froeschlé C. (2003), Detection of Arnold diffusion in Hamiltonian systems, *Physica D*, **182**, 179–187.

Lega E., Froeschlé C. and Guzzo M. (2007), Diffusion in Hamiltonian quasi-integrable systems." In *Lecture Notes in Physics, "Topics in Gravitational Dynamics"*, Benest, Froeschlé, Lega eds., Spinger. In press.

Moser J. (1958), On invariant curves of area-preserving maps of an annulus, *Comm. on Pure and Appl. Math.*, **11**, 81–114.

Murray, N. and Holman, M. (1999), The origin of chaos in the outer solar system. *Science*, **283**, 1877–1881.

Nobili, A.M., Milani, A. and Carpino, M. (1989), Fundamental frequencies and small divisors in the orbits of the outer planets, *Astron. Astrophys.*, **210**, 313–336.

Poincaré H. (1892), Les méthodes nouvelles de la méchanique celeste, Gaulhier–Villars, Paris.

Simo C. (1989), On the analytical and numerical approximation of invariant manifolds, in Modern Methods in Celestial Mechanics, D. Benest, Cl. Froeschlé eds, *Editions Frontières*, 285-329.

Smale S. (1967), Differentiable dynamical systems, *Bulletin of the American Mathematical Society*, **73**, 747-817.

Sussman, G.J. and Wisdom, J. (1988), Numerical evidence that the motion of Pluto is chaotic, *Science*, **241**, 433–437.

Sussman, G.J. and Wisdom, J. (1992), Chaotic evolution of the solar system, *Science*, **257**, 56–62.

Extra-solar multiplanet systems

Sylvio Ferraz-Mello[1], C Beaugé[2] and TA Michtchenko[1]

[1]Instituto de Astronomia, Geofísica e Ciências Atmosféricas, Universidade de São Paulo, Brasil; [2] Observatório Astronómico, Universidad Nacional de Córdoba, Argentina

1 Introduction

This chapter presents some facts about multiplanet systems. By mid-2008, 30 extra-solar systems with 2 or more planets had been identified. Their orbits are shown in Tables 1 and 2. The orbits given in these tables were selected from those published as being the most reliable for dynamical studies. Some additional multiplanet systems have been announced in meetings or otherwise, but are not included in the tables since a document on their orbits is not yet available. Table 2 also shows, for comparison, the two subsystems forming the solar system.

When the ratios of the orbital periods of exoplanets in consecutive orbits are presented in increasing order, in a graph, (Figure 1), we see a more or less continuous distribution between two extreme cases: at the lower end, we have the pair formed by the two Earth-size planets of pulsar PSR B 1257+12, for which $P_2/P_1 \approx 1.5$, and several planet pairs around normal stars (main-sequence and subgiants) with $P_2/P_1 \approx 2$; in the upper end, $P_2/P_1 \gg 100$.

The study of the dynamics of extra-solar planetary systems has revealed an important richness of configurations as can be seen in the recent review by Michtchenko et al. (2007). For the purposes of dynamical studies, these systems may be separated into three main classes of increasing period ratio: (I) planets in close orbits; (II) non-resonant planets with significant secular dynamics; and (III) weakly-interacting planet pairs. In class I, two distinct dynamical situations are possible (see Section 2) and we have to consider two subclasses: (Ia) planets in mean-motion resonance and (Ib) low-eccentricity near-resonant pairs.

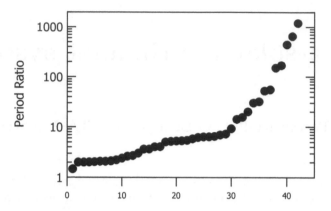

Figure 1. *Ratio of the orbital periods of exoplanets in consecutive orbits. The abscissas are sequence numbers of the data arranged in increasing order.*

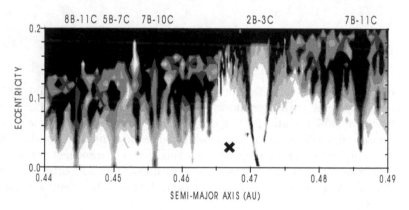

Figure 2. *Dynamical map of the neighborhood of PSR B1257+12 planet C. The main apparent MMRs are indicated on top of the figure. The 'X' shows the actual position of planet C. Gray scale: White: regular (stable) motion; light tones: mild chaos; dark tones: strong chaos; black: Instability generally associated with close approaches. Map constructed using the minimal masses corresponding to $\sin i = 1$. (Reprinted with permission from Ferraz-Mello, S. and Michtchenko, T.A., Extra-solar planetary systems, Revista Mexicana de Astronomía y Astrofísica Series de Conferencias, 14, 7-10, 2002.)*

2   Class I. Planets in close orbits

We put in Class I those planets at the lower end of the distribution shown in Figure 1 orbiting in relatively close orbits (i.e. consecutive orbits with a small period ratio). These planets are subject to strong gravitational interaction and are significantly perturbed on orbital timescales.

A picture of the dynamics in the neighborhood of two planets in close

Table 1. *Extra-Solar Multiplanet Systems - Main List*

Star Star mass	Planet	Mass × sin i (m_{Jup})	Period (days)	Period ratio	a (AU)	e	ω (deg)	T_p (JD - 2,400,000)
GJ 876	d	5.89	1.93776	15:6	0.0208067	0†	0†	52,490.27‡
0.32$M_\odot$	c	0.619	30.340	2:01	0.13030	0.224	198.3	52,494.34‡
	b	1.935	60.940		0.20783	0.025	175.7	52,521.20‡
HD 73526	b	2.07	187.5	2:01	0.66	0.39	172	50,037
1.08$M_\odot$	c	2.30	376.9		1.05	0.40	183	50,184
HD 128311	b	2.18	458.6	2:02	1.10	0.25	110.9	50,210.9
0.84$M_\odot$	c	3.21	928.3		1.76	0.17	195.5	50,012.2
HD 82943	c	1.703	218.7	2:05	0.745	0.361	132.3	51,184.6‡
1.15$M_\odot$	b	1.747	447.5		1.200	0.190	192.0	51,261.5‡
μAra	d	0.032	9.64	32:0	0.091	0.132	112.24	51,119.01
(=HD 160691)	e	0.428	308.59	2:09	0.917	0.061	174.04	51,160.36
1.08$M_\odot$	b	1.611	643.89	5:28	1.498	0.024	42.49	51,112.95
	c	1.586	3403		4.545	0.059	161.78	50,411.91
HD 40307	b	0.0132	4.3115	2:23	0.047	0†	0†	54,562.77
0.77$M_\odot$	c	0.0216	9.620	2:15	0.081	0†	0†	54,551.53
	d	0.0288	20.46		0.134	0†	0†	54,532.42
GJ 581	b	0.0490	5.3687	2:41	0.041	0†	0†	52,999.99
0.31$M_\odot$	c	0.0159	12.931	6:45	0.073	0†	0†	52,996.74
	d	0.0263	83.4		0.25	0†	0†	52,954.1
HD 155358	b	0.89	195.0	2:72	0.628	0.112	162	53,950.0
0.87$M_\odot$	c	0.504	530.3		1.224	0.176	279	54,420.3
55 Cnc	e	0.034	2.81705	5:20	0.038	0.07	248.9	49,999.836
(=ρ^1 Cnc)	b	0.824	14.65162	3:03	0.115	0.014	131.94	50,002.947
0.94$M_\odot$	c	0.169	44.3446	5:86	0.240	0.086	77.9	49,989.34
	f	0.144	260.00	20:1	0.781	0.2	181.1	50,080.91
	d	3.835	5218		5.77	0.025	181.3	52,500.6
HD 69830	b	0.032	8.667	3:64	0.0785	0.10	340	53,496.8
0.86$M_\odot$	c	0.037	31.56	6:24	0.186	0.13	221	53,469.6
	d	0.057	197		0.630	0.07	224	53,358
HD 177830	c	0.186	111.19	3:67	0.514	0.40	100	53,953.7
1.46$M_\odot$	b	1.43	407.88		1.22	0.041	166	50,238
HD 108874	b	1.36	395.4	4:06	1.051	0.07	248.4	50,131.5
0.99$M_\odot$	c	1.018	1605.8		2.68	0.25	17.3	49,584.8
HD 102272	b	5.9	127.58	4:08	0.614	0.05	118	52,146
1.9$M_\odot$	c	2.6	520		1.57	0.68	320	54,135
HD 202206	b	17.5	256.20	5:06	0.83	0.433	161.1	52,175.6
1.15$M_\odot$	c	2.41	1296.8		2.44	0.284	101.8	51,206.4

– Continued

† Fixed arbitrarily; ‡ T_p calculated from given mean anomaly at the dates 52,490.0 (GJ 876) and 51,184.6 (HD 82943);
Note: (Updated in 2008)

orbits is shown in Figure 2. That figure shows the neighborhood of the planet C of the pulsar PSR B1257 +12, which is very close to the orbit of planet B (period ratio close to 3 : 2). This neighborhood is dominated by the resonance 2B–3C ($2n_B - 3n_C \sim 0$) with narrow bands due to several higher-order resonances (indicated above the top axis of Figure 2). That figure shows an almost continuous white (stable) horizontal area along the figure, in very low

Table 1. *Extra-Solar Multiplanet Systems - Main List (Continued)*

Star / Star mass	Planet	Mass × sin i (m_{Jup})	Period (days)	Period ratio	a (AU)	e	ω (deg)	T_p (JD - 2,400,000)
vAnd	b	0.687	4.617113	52.2	0.0595	0.023	63	51,802.64
1.32$M_\odot$	c	1.98	241.23	5.35	0.832	0.262	245.5	50,158.1
	d	3.95	1290.1		2.54	0.258	279	48,827
HD 37124	b	0.61-0.66	~154	~ 5.45	~0.5	<0.25	?	~50,000
0.78$M_\odot$	d	0.6-0.7	~840		~1.6	~0.15	?	49,400-600
	c	0.2-0.7	?		?	?	?	?
HD 12661	b	2.34	262.53	6.39	0.831	0.361	296.3	50,214.1
1.11$M_\odot$	c	1.83	1679		2.86	0.017	38	52,130
HD 74156	b	1.847	51.75	6.56	0.292	0.629	176.45	45,853.54‡
1.24$M_\odot$	d	0.412	339.4	7.29	1.023	0.227	191.81	45,886.3‡
	c	7.995	2475.9		3.848	0.426	262.17	48,129.65‡
47 UMa	b	2.60	1083.2	7.0	2.11	0.049	111	50,173
1.07$M_\odot$	c	1.34	7586		7.73	0.005†	127†	52,134
HD 169830	b	2.88	225.62	9.32	0.81	0.31	148	51,923
1.40$M_\odot$	c	4.04	2102		3.60	0.33	252	52,516
HIP 14810	b	3.91	6.6742	14.8	0.0692	0.1470	158.6	53,694.59
0.99$M_\odot$	c	0.76	95.2914		0.407	0.4088	354.2	53,679.58
HD 168443	b	8.02	58.11289	30.1	0.300	0.5286	172.87	50,047.39
1.05$M_\odot$	c	18.1	1749.5		2.91	0.2125	65.07	50,273.0
HD 11964	c	0.09	38.02	55.5	0.253	0.23	123	52,737
1.12$M_\odot$	b	0.61	2110		3.34	0.06	168	52,290
HD 38529	b	0.852	14.3093	151	0.1313	0.248	91.2	49,991.59
1.47$M_\odot$	c	13.2	2165		3.74	0.3506	15.7	50,085
HD 190360	c	0.057	17.10	169	0.128	<0.1	154	50,000.07
0.96$M_\odot$	b	1.502	2891		3.92	0.36	12.4	50,628.1
HD 217107	b	1.37	7.1269	442	0.074	0.13	21.1	49,998.42
1.05$M_\odot$	c	2.1	3150		4.3	0.55	164	51,030
HD 68988	b	1.92	6.276370	653	0.0704	0.150	40	51,549.73
1.18$M_\odot$	c	5.29	4100		5.32	0.01	129	52,300
HD 187123	b	0.522	3.0965825	1195	0.0426	0.01	5	50,806.65
1.08$M_\odot$	c	1.95	3700		4.80	0.249	250	49,910

‡ T_p calculated from given mean anomaly at the date 45,823.557.

References: Barnes et al. 2008 (HD 74156); Beaugé et al. 2008 (HD 82943); Butler et al. 2006 (v And, HD 12661, HD 11964b); Butler et al. 2008 (HD 177830, HD 68988, HD 187123); Cochran et al. 2007 (HD 155358); Correia et al. 2005 (HD 202206); Fischer et al. 2008 (55 Cnc); Gregory 2007 (HD 11964c); Lovis et al. 2006 (HD 69830); Mayor et al. 2004 (HD 169830); Mayor et al. 2008 (HD 40307); Niedzielski et al. 2008 (HD 102272); Rivera et al. 2005 (GJ 876); Short et al. 2008 (μ Ara); Tinney et al. 2006 (HD 73526); Udry et al. 2007 (GJ 581); Vogt et al. 2005 (HD 128311, HD 108874, HD 37124, HD 190360, HD 217107); Wittenmyer et al. 2005 (47 UMa); Wright et al. 2007 (HIP 14810, HD 168443).

eccentricities. Therefore, one pair of planets with small period ratio orbits may be stable for a wide domain of semi-major axes.[1]

This figure shows the main stability conditions in this case: (1) Planets in mean-motion resonance (MMR), able to have larger eccentricities; (2) Planets in almost circular orbits (i.e. very low eccentricities). They are considered as forming the two subclasses Ia and Ib discussed below.

---

[1] Figure 2 also shows that the only robust (wide) domain where the motion is not affected by stronger chaos at larger eccentricities lies inside the 3/2 resonance.

Table 2. *Other Planetary Systems*

Far away planetary systems[†]							
Star	Planet	Mass	Period	Period	Distance[‡] to		
Star mass		(m_{Jupiter})	(years)	ratio	Star (AU)		
OGLE-06-109L	b	0.71	5.00	2.8	2.3		
0.5 $M_\odot$	c	0.27	13.96		4.6		

Pulsar planetary system							
Pulsar	Planet	Mass	Period	Period	Semi-major	Eccentri-	Inclina-
Pulsar mass		(m_{Earth})	(days)	ratio	axis (AU)	city	tion[††]
PSR B1257+12	A	0.02[‡‡]	25.262	2.63	0.19	0.	
1.4$M_\odot$(adopt.)	B	4.3	66.5419	1.47	0.36	0.0186	53°
	C	3.9	98.2114		0.46	0.0252	47°

Outer Solar System						
Star	Planet	Mass	Period	Period	Semi-major	Eccentri-
		(m_{Jupiter})	(years)	ratio	axis (AU)	city
Sun	Jupiter	1.	11.866	2.500	5.204	0.0489
	Saturn	0.299	29.668	2.831	9.584	0.0570
	Uranus	0.0457	83.987	1.958	19.178	0.0468
	Neptune	0.0540	164.493		30.006	0.0112

Inner Solar System						
Star	Planet	Mass	Period	Period	Semi-major	Eccentri-
		(m_{Earth})	(days)	ratio	axis (AU)	city
Sun	Mercury	0.0539	88.06	2.554	0.3871	0.2056
	Venus	0.7953	224.94	1.623	0.7233	0.0068
	Earth	1.[§]	365.25	1.882	1.0000	0.0167
	Mars	0.1048	687.75		1.5237	0.0934

[†]Discovered by microlensing. Distance to Sun ∼ 1.5 kpc. [‡]Projected on the tangent plane to the celestial sphere. [††]Over the tangent plane to the celestial sphere. [‡‡]Adopting the inclination $i = 50°$. [§] Moon mass not included; $M_{\text{Jupiter}} = 310.22 M_{\text{Earth}}$.
References: Gaudi et al. 2008 (OGLE-2006-BGL-109L); Konacki and Wolszczan, 2003 (PSR B1257+12).
Note: (Updated in 2008)

2.1 Class Ia. Planets in resonant orbits (MMR)

These are among the more interesting extra-solar systems for celestial mechanics studies. Systems such as GJ 876 (= Gliese 876) and HD 82943 have been the subject of several detailed studies (e.g. Beaugé and Michtchenko, 2003; Beaugé et al. 2006, 2008; Ferraz-Mello et al. 2005; Goździevski and Konacki, 2006; Laughlin et al. 2005; Lee et al. 2006; Psychoyos and Hadjidemetriou, 2005; Rivera et al. 2005). The most important feature in these systems is the interior resonance characterized by the capture of the critical angle[2]

$$\theta_1 = 2\lambda_2 - \lambda_1 - \varpi_1$$

about 0 or π. (For a complete dynamical study of planets in 2:1 mean-motion resonance, see Michtchenko et al. 2008a,b.)

In the well-known case of asteroids trapped into a 2:1 resonance, the pericenter of the asteroid orbit (whose longitude is ϖ_1) continues to rotate. That is, $\Delta\varpi = \varpi_1 - \varpi_2$ has a monotonic time variation. However, in the case of two planets, it may happen that both θ_1 and

$$\theta_2 = 2\lambda_2 - \lambda_1 - \varpi_2$$

---

[2]λ_i and ϖ_i are the mean longitudes and longitudes of the pericenters of the two planets; in the case of asteroids the subscript $i = 2$ refers to Jupiter.

acquire an oscillatory motion around 0 or π. Consequently, in that case, $\Delta\varpi$ is no longer circulating but oscillating around a fixed value (Ferraz-Mello et al. 2003). This is the same phenomenon known as corotation resonance in disc and ring dynamics (motion corotating with the resonant patterns of the potential), extended to beyond the particular 1:1 resonance case of the epicyclic orbits theory. The characterization of corotation resonance given by Greenberg and Brahic (1984): 'resonance that depends on the eccentricity of the perturbing satellite, rather than on the eccentricity of the perturbed particle' means that we have a corotation resonance when θ_1 and θ_2 are both oscillating around fixed values. A stationary solution with constant θ_1, θ_2 and $\Delta\varpi$ is called 'apsidal corotation resonance' (ACR).

The resonant planar planetary three-body problem (averaged over short-period terms) is a system with two degrees of freedom. ACRs are thus solutions for which the averaged angles θ_1 and θ_2 and the averages of the momenta I_1, I_2, conjugated to them, remain constant in time. It is important to notice that these equilibrium solutions of the averaged equations correspond to periodic orbits of the non-averaged problem in a rotating reference frame (Hadjidemetriou, 2006, 2008).

Although ACRs have gained certain notoriety in exoplanetary dynamics, they are not new and can also be found in our own solar system. It has long been known (see Greenberg 1987 and references therein) that the Io-Europa pair of satellites is trapped in a 2/1 ACR: θ_1 oscillates around 0 and $\Delta\varpi$ oscillates around π (the pericenters are *anti-aligned*). If the nomenclature introduced by Beaugé et al (2006) is used, we say that the Io-Europa configuration is in a $(0, \pi)$-ACR, a denomination more accurate than just saying that the pericenters are *anti-aligned*.

The state of motion of the two planets of GJ 876 is a $(0,0)$-ACR: the angular variables oscillate around $\theta_1 = 0$ and $\Delta\varpi = 0$ (the pericenters are *aligned*).

The difference in behavior in these corotations is associated with the eccentricities of these solutions (see Lee and Peale, 2002; Beaugé et al. 2003): The $(0, \pi)$ solutions occur for low eccentricities of the bodies, while $(0, 0)$ corotations occur for larger eccentricities.

The first study of families of periodic solutions of the exact equations of planets in 2/1 MMR, considering different eccentricities but constant planetary masses, is due to Hadjidemetriou (2002). By means of numerical continuation of initially circular orbits of planets with masses similar to GJ 876 b and c, Hadjidemetriou (2002) found that the $(0, \pi)$ and $(0, 0)$ families are actually linked at $(e_1, e_2) = (0.097, 0)$, and it is possible to pass from one to another by a smooth variation of the total angular momentum.

Systematic searches resulted in the construction of catalogues of stable and unstable apsidal corotation resonances for the resonant planar planetary three-body problem, including both symmetric and asymmetric solutions (Beaugé et al., 2006; Michtchenko et al. 2006). In these references, the calculations were performed using a new approach based on the numerical determination of the averaged Hamiltonian function and their extremals, yielding precise results for any values of the eccentricities and semi-major axes with the exception of the immediate vicinity of the collision curve. They include results for the 2/1, 3/2, 3/1, and 4/1 mean-motion resonances; the 5/1 and 5/2 commensurabilities were also briefly discussed. Using the exact equations of motion (i.e. without averaging) the 2/1 and 3/1 families were also thoroughly studied by Voyatzis and Hadjidemetriou (2005, 2006). It is worth noting that, up to second order of the masses, the position of corotations in the e_1, e_2-plane is only a function of the mass ratio m_2/m_1 of the planets, not depending on the individual masses themselves (provided that they are not large).

In addition to the resonant planets whose period ratio is close to 2, we have to mention two other systems. The first one is HD 202206 whose period ratio is close to 5. These orbits are not so close one to another as in the other cases, but the gravitational interaction between the two planets is enhanced by the fact that the planets in this system are very large (one of them is indeed a brown dwarf star with minimal mass 17.5 $M_\odot$). The other system is formed by the planets b, c of 55 Cnc. These two planets were believed to be in a 3/1 resonance up to very recently and several dynamical studies showed the likelihood of such solution (Ferraz-Mello et al. 2003; Kley, 2003; Zhou et al. 2004, 2008; Voyatzis and Hadjidemetriou, 2006). The more recent orbit determination (Fischer et al. 2008) confirms the almost commensurability of the two periods (period ratio 3.03) but fails to show the libration of the critical angles associated with $3\lambda_2 - \lambda_1$. However, the uncertainty in the determination of eccentricities and pericenter longitudes is important and small changes in the data set are often enough to change the behavior of one given solution (see Beaugé et al. 2008).

2.2   Class Ib. Low-eccentricity near-resonant pairs

The second general stability condition for planets in close orbits is small eccentricity. The best example in this case is the solar system where the period ratios of consecutive planets are small, ranging between 1.6 and 2.8 (with exception of the pair Mars-Jupiter, of course). Solar-system planets are not in mean-motion resonance (see Table 2) and are known to be stable for many billions of years. The stability of the outer solar system has been the subject of many studies. We cite the exploration of a large portion of the phase space where the outer solar system evolves and the construction of dynamical maps of the regions around the Jovian planets (Michtchenko and Ferraz-Mello, 2001). That study has shown that these regions are densely filled by two- and three-planet mean-motion resonances that may generate instabilities in planetary motion. The actual position of the giant planets is however far off these instabilities. The dynamical map in that case shows similarity with the map shown in Figure 2, but is much more complex because we have, in the outer solar system, 4 giant planets in relatively close orbits.

Several of the exoplanet systems shown in Table 1 may belong to this class.

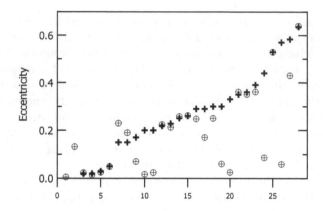

Figure 3. *Distribution of the eccentricities of the exoplanets considered in Ferraz-Mello et al. 2005, shown in increasing order (crosses), compared to the new values given in Table 1 (open circles). The planets of HD 202206 and HD 169830, whose orbits were not updated recently, were not included. The significant decrease of the eccentricities of several exoplanets is notorious.*

The low-eccentricity planets of 47UMa were for a long time believed to belong to this class. Several of the preliminary orbits have shown period ratios close to 5/2 and to 8/3, but these near commensurabilities have not been confirmed by new observations. On the contrary, observations over a time span of about 20 years (including many pre-discovery measurements) show that the previously suspected 8-yr orbital period of the outer planet may be in fact much longer (Wittenmyer et al. 2005) and the period ratio of the best solutions is rather close to 7.0.

The planets b, c of 55 Cnc, for a long time considered as an example of class Ia systems, are now more likely in class Ib. The orbits are very close to the 3/1 commensurability but the best-fit eccentricities are small (0.014 and 0.086) and the oscillation of the critical angle is not observed in the simulations done using the best-fit orbital data.

One may note that the eccentricities of the exoplanets in close orbits with period ratio less than 3, with the exception of the 4 first ones shown in Table 1, are rather small. For large period ratios we see some larger eccentricities but we have to wait for new observations before confirming their values. We may note that the comparison of the current eccentricities to those appearing in the tables used in our previous analysis (Ferraz-Mello et al. 2005), done 4 years ago, shows a notorious decrease of several of the values formerly used (see Figure 3) leading us to expect similar decreases in the forthcoming orbit determinations.

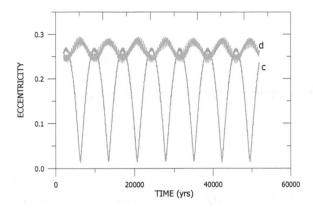

Figure 4. *Variation of the eccentricities of the planets υ And c and υ And d.*

3 Class II. Non-resonant planets with significant secular dynamics

The period ratios of the planet pairs considered in this class lie above ∼ 4 and are large enough to make more difficult a capture into one MMR (with the already discussed exception of the massive planets of HD 202256). The gravitational interaction between these planets may be strong, but the conservation of angular momentum limits the eccentricity variations, allowing the system to remain stable. Being far off a MMR and assuming the absence of other perturbations, the semi-major axes remain nearly constant and the angular momentum conservation of two coplanar planets allows us to write

$$\text{AMD} = m_1 n_1 a_1^2 (1 - \sqrt{1 - e_1^2}) + m_2 n_2 a_2^2 (1 - \sqrt{1 - e_2^2}). \tag{1}$$

The Angular Momentum Deficit or AMD (Laskar 2000), is a constant formed by the sum of two positive definite terms. Therefore, when one of the terms grows, the other may necessarily decrease so as to keep their sum constant. Since the only quantities allowed to vary are the eccentricities, the variation of the eccentricities will be limited. For example, if $e_1 \to 0$, then AMD $\to$ $m_2 n_2 a_2^2 (1 - \sqrt{1 - e_2^2})$. Since AMD is a given constant, this sets a limiting value for the second term. If this limit corresponds to an eccentricity $e_2 < 1$, it sets a limiting eccentricity value that cannot be exceeded since there is no value of e_1 allowing it to exceed this limit as long as the AMD is kept constant. (Indeed, if the AMD of a system is too large and the considered limit corresponds to an eccentricity $e_2 > 1$, we have to use the properties of the AMD. Otherwise, neither eccentricity can go below a lower limit, without having the other going beyond 1 and the system being disrupted.) Figure 4 shows the behavior of the eccentricities of the planets of υ And c and d.

Another important feature of the planets in class II is the consequence of the secular dynamics on the behavior of the pericenters. This can be eas-

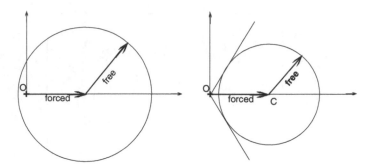

Figure 5. *Vector composition of eccentricities of two planets in Laplace-Lagrange secular theory. (Left):* $e_{\text{free}} > e_{\text{forced}}$; *(right):* $e_{\text{free}} < e_{\text{forced}}$.

ily seen if we recall some elementary results issued from the linear Laplace-Lagrange theory. For a thorough presentation of it, see Pauwels (1983). The gravitational interaction of two planets in near-circular orbits introduces a main perturbation in eccentricity and pericenter of the planets which is a constant forced eccentricity, in addition to the free eccentricity. The forced eccentricity is directly proportional to the strength of the interaction of the two planets. If we represent the result in polar coordinates, using one eccentricity as the radius vector and the relative pericenter angle $\Delta\varpi = \varpi_1 - \varpi_2$ as the polar angle, the forced eccentricity is given by a constant vector on the axis $\Delta\varpi = 0$ (mod π). In addition to the forced eccentricity, there is the free eccentricity of the orbit. In the polar diagram, the free eccentricity is represented by a vector constant in modulus but rotating, since the mutual perturbation of the two planets forces the orbits to precess in their planes. The eccentricity of the planet is given by the modulus of the sum of these two vectors and the relative pericenter angle $\Delta\varpi$ is the position angle of the vector sum.

Figure 5 shows that, in absence of the free eccentricity, the pericenter of the two planets would appear always locked in the direction of the forced eccentricity. However, in general, the free eccentricity is not damped to zero and the resulting eccentricity and pericenters are given by the sum of the two vectors. The kinematical behaviour of the pericenters is determined by the values of the two eccentricities. If $e_{\text{free}} < e_{\text{forced}}$, (Figure 5 right), the relative pericenter angle $\Delta\varpi$ is limited by the angles defined by the two tangents to the circle passing by O. The resulting motion is the oscillation of the pericenters around the exact alignment (or anti-alignment, in other examples) and the variation of the eccentricity between the extremals is given by $e_{\text{forced}} - e_{\text{free}}$ and $e_{\text{forced}} + e_{\text{free}} < 1$. This situation is usually called a 'secular resonance'. If the free eccentricity is large enough, so that $e_{\text{free}} > e_{\text{forced}}$ (Figure 5 left), the relative pericenter angle $\Delta\varpi$ is no longer limited and can take all values between 0 and 360 degrees. The resulting motion is the unconstrained circulation of the pericenters and an oscillation of the eccentricity between the extremals given by $e_{free} - e_{forced}$ and $e_{forced} + e_{free} < 1$. Nonlinear theories lead to phenomena much more complex than in linear theory but, in general, showing also the overall feature above described (see Michtchenko and Malhotra, 2004).

The kinematics of the two cases considered in Figure 5 are different since in one case the two pericenters are oscillating one about another, while in the other case their relative motion is a circulation. However, this difference is just due to a quantitative circumstance and has no dynamical significance. There is no dynamical separatrix and to class the transition from one case to another as a 'separatrix crossing' is not justified. In planetary systems of classes Ib and II, the 'secular resonance' is generally due to the fact that the forced eccentricities are large because of the closeness of the two orbits.

Amongst the most conspicuous examples in this class we find four important systems: the outermost planets of μ Ara, HD 12661, the outer planets of υ And and, now, 47 UMa. In all cases, they do not seem to be in MMR or close to one; the outer planets of υ And are paradigms of systems showing apsidal lock due to non-resonant secular dynamics.

We place in class II those cases in which the gravitational interaction of the two planets is strong enough to create a large forced eccentricity. As a consequence, these cases are characterized by the apsidal lock of the pericenters.

4   Class III. Weakly-interacting planet pairs

These are the planets given at the bottom of Table 1. One important example is HD 38529 where $P_2/P_1 \sim 150$. The large period ratios (> 30) allow these planets to be considered as weakly interacting planets; the mutual gravitational interaction exists, but is less important than in the previous case and the probability of capture in a MMR is negligible. We recall that in many dynamical studies of the outer solar system, the mass of the inner planets is just added to the mass of the Sun!

Two of the candidates in this category are the planets around HD 168443 and HD 74156. However, these systems have some very massive planets and the hierarchical models should be used to study these pairs prior to specific analyses.

The only dynamical difference between these planets and those in class II is that, in this case, the gravitational interaction of the two planets is weak and, as a consequence, the forced eccentricity is small and they are characterized by the relative circulation of the two pericenters.

One important point to take into consideration is that the large period ratio in many cases may just be due to the fact that the observations were not yet sufficient to show additional planets in these systems, which may exist between the two currently known planets. This happened to half of the planets classed as hierarchical 4 years ago, and we may expect a similar fate for many of those appearing as hierarchical now. In addition, we may expect that many of the high eccentricities now obtained for these planets will decrease when more observations are added and new planets discovered in these systems.

5 Equations of motion for N planets

In order to study the motions in a system of planets, we have to consider their mutual gravitational interaction. Usually, this study is done considering the star and the planets as mass-points and forces given by Newton's law ($N + 1$-body models). However, one must keep in mind that in several circumstances this may be not enough. We may easily devise several circumstances in which the consideration of additional forces is necessary. As examples we mention (a) the case of hot Jupiters, in which the proximity of the planet to the star makes necessary to consider the tidal deformation of their bodies (see, e.g. Ferraz-Mello et al. 2008); (b) the motion of resonant planets in low orbits: in this case the relativistic motion of the pericenter may affect the behavior of the critical angles θ_i (see Section 2.1); (c) the case of planets of a star belonging to a pair of stars in which the second star is close enough to disturb the motion of the planets, etc.

The barycentric equations of the N-planet problem are obtained using the basic principles of mechanics. We may either use Newton's laws or the classical formulations of Lagrangian or Hamiltonian mechanics. Because of the theoretical advantages of the Hamiltonian formulation, we will adopt it, but the first set of equations discussed in this section may easily be obtained using Newton's laws.

We denote as $\mathbf{X}_i$ ($i = 0, 1, \cdots, N$) the position vectors of the N+1 bodies (the star and the N planets) with respect to an inertial system, and as $\mathbf{p}_i = m_i \dot{\mathbf{X}}_i$ their linear momenta. Since the kinetic energy is a homogeneous quadratic form in the velocities, the Hamiltonian of the system is given simply by the sum of the kinetic and potential energies:

$$H = T + U = \frac{1}{2} \sum_{i=0}^{N} \frac{\mathbf{p}_i^2}{m_i} - \sum_{i=0}^{N} \sum_{j=i+1}^{N} \frac{Gm_i m_j}{\Delta_{ij}} \qquad (2)$$

where G is the constant of gravitation and $\Delta_{ij} = |\mathbf{X}_i - \mathbf{X}_j|$. The system has $6(N + 1)$ equations, that is, $3(N + 1)$ degrees of freedom.

The variables $\mathbf{X}_i$ and $\mathbf{p}_i$ are canonical and the resulting system of equations is

$$\frac{d\mathbf{X}_i}{dt} = \nabla_{\mathbf{p}_i} H = \frac{\mathbf{p}_i}{m_i} \qquad (i = 0, 1, 2, \cdots, N) \qquad (3)$$

$$\frac{d\mathbf{p}_i}{dt} = -\nabla_{\mathbf{X}_i} H = -\sum_{j=0}^{N} {}^{*} \frac{Gm_i m_j}{\Delta_{ij}^3} (\mathbf{X}_i - \mathbf{X}_j). \qquad (4)$$

where an asterisk is used in $\sum^*$ to remind that the term $j = i$ is not considered in the summation. These equations may be written as the well-known $N + 1$ second-order equations

$$\frac{d^2\mathbf{X}_i}{dt^2} = -\sum_{j=0}^{N}{}^* \frac{Gm_j(\mathbf{X}_i - \mathbf{X}_j)}{\Delta_{ij}^3}. \qquad (i = 0, 1, 2, \cdots, N) \quad (5)$$

This equation is nothing but the expression of the Newton's law in the form acceleration=force/mass for each body in the problem under consideration.

5.1 Astrocentric equations of motion

Eq. (5) can be easily reduced to $6N$ equations if we introduce the astrocentric position vectors:

$$\mathbf{r}_i = \mathbf{X}_i - \mathbf{X}_0. \qquad (6)$$

Combining Eq. (5) for the subscripts $i \neq 0$ and 0, we obtain

$$\begin{aligned}
\frac{d^2\mathbf{r}_i}{dt^2} &= \frac{d^2\mathbf{X}_i}{dt^2} - \frac{d^2\mathbf{X}_0}{dt^2} \\
&= -\frac{G(m_0 + m_i)\mathbf{r}_i}{r_i^3} - \sum_{j=1}^{N}{}^* Gm_j \left(\frac{\mathbf{r}_i - \mathbf{r}_j}{\Delta_{ij}^3} + \frac{\mathbf{r}_j}{r_j^3} \right),
\end{aligned} \qquad (7)$$

where we have introduced the astrocentric distances $r_i = \Delta_{i0}$.

Since there are N vector variables $\mathbf{r}_i$, the system is reduced to order $6N$. The 6 variables that disappeared from the equations may be retrieved if necessary using the barycentric inertial property $\sum_{i=0}^{N} m_i\dot{\mathbf{X}}_i = \text{const}$, and its time integral.

These equations are exact and they are the preferred ones in studies of the motion of the planets by means of numerical simulations.

6 Reduction of the Hamiltonian equations

The barycentric equations were reduced to 3N degrees of freedom through the introduction of astrocentric position vectors and an intrinsic use of the conservation laws giving the inertial motion of the barycenter. However, it is not trivially possible to transform Eq. (7) into a canonical system of equations. More specifically, the system of variables formed by the relative position vectors $\mathbf{r}_i$ and momenta $\mathbf{p}_i = m_i\dot{\mathbf{r}}_i$ is not canonical.

There are, however, two sets of variables which are canonical and allow the given Hamiltonian system to be reduced to 3N degrees of freedom. The most popular reduction, due to Jacobi, is widely used in planetary studies. The less popular reduction is due to Poincaré. It was first published in 1897, but Poincaré himself did not use it because of difficulties related with the definition of the associated Keplerian elements (see Poincaré, 1905). It appeared in the literature from time to time and started being used more frequently around the 1980s (Yuasa and Hori, 1979; Hori, 1985; Laskar, 1990). Hagihara (1970) says that it was discovered by Cauchy.

6.1 Jacobi's canonical coordinates

In these coordinates, the position and velocity of the planet P_1 are given in a reference frame with origin in P_0; the position and velocity of P_2 are given in a reference frame with origin at the barycenter of the system formed by P_0 and P_1; the position and velocity of P_3 are given in a reference frame with origin at the barycenter of the system formed by P_0, P_1 and P_2; and so on. If we denote with $\boldsymbol{\varrho}_i$ $(i = 1, \cdots, N)$ the position vectors thus defined, we have

$$
\begin{aligned}
\boldsymbol{\varrho}_1 &= \mathbf{X}_1 - \mathbf{X}_0 \\
\boldsymbol{\varrho}_2 &= \mathbf{X}_2 - \frac{m_0\mathbf{X}_0 + m_1\mathbf{X}_1}{\sigma_1} \\
\boldsymbol{\varrho}_3 &= \mathbf{X}_3 - \frac{m_0\mathbf{X}_0 + m_1\mathbf{X}_1 + m_2\mathbf{X}_2}{\sigma_2} \\
&\qquad\qquad \cdots \\
\boldsymbol{\varrho}_N &= \mathbf{X}_N - \frac{m_0\mathbf{X}_0 + m_1\mathbf{X}_1 + \cdots + m_{N-1}\mathbf{X}_{N-1}}{\sigma_{N-1}}
\end{aligned}
\tag{8}
$$

where, for sake of simplicity, we have introduced the partial sums

$$
\sigma_k = m_0 + m_1 + \cdots + m_k.
\tag{9}
$$

The inverses are

$$
\begin{aligned}
\mathbf{X}_1 - \mathbf{X}_0 &= \boldsymbol{\varrho}_1 \\
\mathbf{X}_2 - \mathbf{X}_0 &= \frac{m_1\boldsymbol{\varrho}_1}{\sigma_1} + \boldsymbol{\varrho}_2 \\
\mathbf{X}_3 - \mathbf{X}_0 &= \frac{m_1\boldsymbol{\varrho}_1}{\sigma_1} + \frac{m_2\boldsymbol{\varrho}_2}{\sigma_2} + \boldsymbol{\varrho}_3. \\
&\qquad\qquad \cdots \\
\mathbf{X}_N - \mathbf{X}_0 &= \frac{m_1\boldsymbol{\varrho}_1}{\sigma_1} + \frac{m_2\boldsymbol{\varrho}_2}{\sigma_2} + \cdots + \frac{m_{N-1}\boldsymbol{\varrho}_{N-1}}{\sigma_{N-1}} + \boldsymbol{\varrho}_N.
\end{aligned}
\tag{10}
$$

For a generic subscript, we have

$$
\mathbf{X}_k - \mathbf{X}_0 = \boldsymbol{\varrho}_k + \sum_{j=1}^{k-1} \frac{m_j}{\sigma_j}\boldsymbol{\varrho}_j.
\tag{11}
$$

To complete the set of vectors describing the positions of the N+1 bodies, we introduce:

$$
\boldsymbol{\varrho}_0 = \frac{m_0\mathbf{X}_0 + m_1\mathbf{X}_1 + m_2\mathbf{X}_2 + \cdots + m_N\mathbf{X}_N}{\sigma_N}.
\tag{12}
$$

To have a system of canonical variables equivalent to that of the barycentric formulation, we have to introduce the new momenta $\boldsymbol{\pi}_i$ $(i = 0, 1, \cdots, N)$. This can be easily done by constructing the generating function $S(\mathbf{X}, \boldsymbol{\pi}) = \sum_0^N \int \boldsymbol{\varrho}_i d\boldsymbol{\pi}_i$ and obtaining the new momenta by means of the differentiation $\mathbf{p}_i = \nabla_{\mathbf{X}_i} S(\mathbf{X}, \boldsymbol{\pi})$ (see Ferraz-Mello, 2007, Sec. 1.3). Hence

$$\mathbf{p}_1 = \boldsymbol{\pi}_1 - \frac{m_1 \boldsymbol{\pi}_2}{\sigma_1} - \frac{m_1 \boldsymbol{\pi}_3}{\sigma_2} - \cdots + \frac{m_1 \boldsymbol{\pi}_0}{\sigma_N}$$

$$\mathbf{p}_2 = \boldsymbol{\pi}_2 - \frac{m_2 \boldsymbol{\pi}_3}{\sigma_2} - \cdots + \frac{m_2 \boldsymbol{\pi}_0}{\sigma_N}$$

$$\mathbf{p}_3 = \boldsymbol{\pi}_3 - \frac{m_3 \boldsymbol{\pi}_4}{\sigma_3} - \cdots + \frac{m_3 \boldsymbol{\pi}_0}{\sigma_N} \tag{13}$$

$$\cdots$$

$$\mathbf{p}_N = \boldsymbol{\pi}_N + \frac{m_N \boldsymbol{\pi}_0}{\sigma_N}$$

and

$$\mathbf{p}_0 = -\boldsymbol{\pi}_1 - \frac{m_0 \boldsymbol{\pi}_2}{\sigma_1} - \frac{m_0 \boldsymbol{\pi}_3}{\sigma_2} - \cdots + \frac{m_0 \boldsymbol{\pi}_0}{\sigma_N} \tag{14}$$

6.1.1 The equations

A lengthy, but simple calculation shows that the total kinetic energy, with the new variables, is

$$T = \frac{\pi_0^2}{2\sigma_N} + \sum_{k=1}^N \frac{\pi_k^2}{2\tilde{\beta}_k} \tag{15}$$

where $\tilde{\beta}_k$ are the so-called reduced masses of the Jacobian formulation, defined by

$$\tilde{\beta}_k = \frac{m_k \sigma_{k-1}}{\sigma_k}. \tag{16}$$

Hence

$$H = \frac{\pi_0^2}{2\sigma_N} + \sum_{k=1}^N \frac{\pi_k^2}{2\tilde{\beta}_k} + U. \tag{17}$$

The reduction of the system is, now, immediate. We may note, beforehand, that the distances Δ_{ij} depend only on the variables $\boldsymbol{\varrho}_1, \boldsymbol{\varrho}_2, \cdots, \boldsymbol{\varrho}_N$ and, therefore, the variable $\boldsymbol{\varrho}_0$ is ignorable. Consequently, $\boldsymbol{\pi}_0$ is a constant. The resulting equations may be separated into two parts:

A. The first pair of equations, corresponding to the subscript 0,

$$\dot{\boldsymbol{\pi}}_0 = 0 \qquad\qquad \dot{\boldsymbol{\varrho}}_0 = \nabla_{\boldsymbol{\pi}_0} H = \frac{\boldsymbol{\pi}_0}{\sigma_N}; \tag{18}$$

B. The canonical equations in the variables ϱ_i, π_i $(i \neq 0)$ given by the reduced Hamiltonian

$$H = \sum_{k=1}^{N} \frac{\pi_k^2}{2\tilde{\beta}_k} + U. \tag{19}$$

This separated subsystem has 3N degrees of freedom.

Let us write the Hamiltonian of the reduced system as $H = H_0 + R$. For that purpose, the potential energy is separated in two parts as $U = U_0 + R$, the first one of order $\mathcal{O}(m_i)$ and the second one of order $\mathcal{O}(m_i^2)$. The terms of order $\mathcal{O}(m_i)$ in U are $\frac{Gm_0 m_k}{\Delta_{0k}}$ $(k = 1, 2, \cdots, N)$. However, this separation is somewhat arbitrary and we can replace, in the main term, Δ_{0k} by $\rho_k = |\varrho_k|$ and m_0 by σ_{k-1}. The resulting main term differs from the previous one by quantities of order $\mathcal{O}(m_i^2)$; this difference may be added to R in order to preserve the total U unchanged. There results:

$$U_0 = -G \sum_{k=1}^{N} \frac{\sigma_{k-1} m_k}{\rho_k} \tag{20}$$

and

$$R = -G \sum_{i=1}^{N} \sum_{j=i+1}^{N} \frac{m_i m_j}{\Delta_{ij}} - G \sum_{k=1}^{N} m_k \left(\frac{m_0}{\Delta_{0k}} - \frac{\sigma_{k-1}}{\rho_k} \right) \tag{21}$$

(See Poincaré, 1905, §31).

The terms T and U_0 define the Hamiltonian

$$H_0 = \sum_{k=1}^{N} \left(\frac{\pi_k^2}{2\tilde{\beta}_k} - G \frac{\sigma_{k-1} m_k}{\rho_k} \right) = \sum_{k=1}^{N} \left(\frac{\pi_k^2}{2\tilde{\beta}_k} - \frac{G\sigma_k \tilde{\beta}_k}{\rho_k} \right). \tag{22}$$

Each term

$$F_k = \frac{\pi_k^2}{2\tilde{\beta}_k} - \frac{\tilde{\mu}_k \tilde{\beta}_k}{\rho_k} \tag{23}$$

where

$$\tilde{\mu}_k = G\sigma_k; \tag{24}$$

is the Hamiltonian of a two-body problem in which the mass-point m_k, is moving around one point with mass $\sigma_{k-1} = m_0 + m_1 + \cdots + m_{k-1}$, lying at the center of gravity of the first k mass-points. Indeed, from the last Hamiltonian, we obtain the second-order differential equation

$$\ddot{\varrho}_k = -\tilde{\mu}_k \frac{\varrho_k}{\rho_k^3} = -G(\sigma_{k-1} + m_k) \frac{\varrho_k}{\rho_k^3}. \tag{25}$$

6.2 Poincaré's relative canonical coordinates

In Poincaré's reduction to N degrees of freedom, the variables are the components of the astrocentric position vectors $\mathbf{X}_i - \mathbf{X}_0$ and the momenta are the same momenta $\mathbf{p}_i$ of the inertial formulation when the inertial system is centered at the barycenter of the system of $N+1$ bodies. Hence,

$$\mathbf{r}_i = \mathbf{X}_i - \mathbf{X}_0, \qquad\qquad \mathbf{\Pi}_i = \mathbf{p}_i. \qquad (26)$$

The given system has N+1 bodies and, therefore, we need to introduce one more pair of (vector) variables. Let them be

$$\mathbf{r}_0 = \mathbf{X}_0, \qquad\qquad \mathbf{\Pi}_0 = \sum_{i=0}^{N} \mathbf{p}_i. \qquad (27)$$

A trivial calculation using the simple canonical condition $\sum_0^N (\mathbf{\Pi}_i d\mathbf{r}_i - \mathbf{p}_i d\mathbf{X}_i) = 0$ shows that the variables $(\mathbf{r}_i, \mathbf{\Pi}_i)$ $(i = 0, 1, \cdots, N)$ are canonical.

Let us, now, write the Hamiltonian in terms of the new variables. The transformations of T and U give, respectively,

$$T = \frac{1}{2} \sum_{i=1}^{N} \frac{\Pi_i^2}{m_i} + \frac{1}{2} \sum_{i=1}^{N} \frac{\Pi_i^2}{m_0} + \frac{1}{2} \frac{\Pi_0^2}{m_0} - \sum_{i=1}^{N} \frac{\mathbf{\Pi}_0 \cdot \mathbf{\Pi}_i}{m_0} + \sum_{i=1}^{N} \sum_{j=i+1}^{N} \frac{\mathbf{\Pi}_i \cdot \mathbf{\Pi}_j}{m_0} \qquad (28)$$

and

$$U = -G \sum_{i=1}^{N} \frac{m_0 m_i}{r_i} - G \sum_{i=1}^{N} \sum_{j=i+1}^{N} \frac{m_i m_j}{\Delta_{ij}} \qquad (29)$$

where $\Pi_i = |\mathbf{\Pi}_i|$ and $r_i = |\mathbf{r}_i| = |\Delta_{0i}|$.

The reduction of the system is also immediate. We note, beforehand, that the variable $\mathbf{r}_0$ is ignorable. Consequently, $\mathbf{\Pi}_0$ is a constant. (By construction, $\mathbf{\Pi}_0 = 0$.) The resulting equations may be separated into two parts:

A. The first pair of equations, corresponding to the subscript 0:

$$\dot{\mathbf{\Pi}}_0 = 0 \qquad\qquad \dot{\mathbf{r}}_0 = \nabla_{\mathbf{\Pi}_0} H = \dot{\mathbf{r}}_0 = -\sum_{i=1}^{N} \frac{\mathbf{\Pi}_i}{m_0}. \qquad (30)$$

B. The canonical equations in the variables $(\mathbf{r}_i, \mathbf{\Pi}_i)$ given by H, which form a subsystem with 3N degrees of freedom separated from the previous one.

Let us write the Hamiltonian of the reduced system as $H = H_0 + R$ where

$$H_0 = \sum_{i=1}^{N} \left(\frac{1}{2} \frac{p_i^2}{\beta_i} - \frac{\mu_i \beta_i}{r_i} \right) \qquad (31)$$

$$R = \sum_{i=1}^{i=N} \sum_{j=i+1}^{j=N} \left(-\frac{Gm_i m_j}{\Delta_{ij}} + \frac{\mathbf{p}_i \cdot \mathbf{p}_j}{m_0} \right) \tag{32}$$

and

$$\mu_i = G(m_0 + m_i) \qquad\qquad \beta_i = \frac{m_0 m_i}{m_0 + m_i}. \tag{33}$$

We went back to the notation $\mathbf{p}_i$ for the momenta, instead of $\mathbf{\Pi}_i$, since both denote the same quantity when the subscript $i = 0$ is not considered.

We also note that H_0 is of the order of the planetary masses m_i while R is of order two with respect to these masses. Then H_0 may be seen as the new expression for the undisturbed energy while R is the potential energy of the interaction between the planets.

It is interesting to conclude this section in the same way as the previous one. Each term

$$F_k = \frac{1}{2} \frac{\mathbf{p}_k^2}{\beta_k} - \frac{\mu_k \beta_k}{r_k} \tag{34}$$

is the Hamiltonian of a two-body problem in which the mass-point m_k is moving around the mass point m_0. In fact, from the last Hamiltonian it is easy to obtain the second-order differential equation

$$\ddot{\mathbf{r}}_k = -\mu_k \frac{\mathbf{r}_k}{r_k^3} = -G(m_0 + m_k)\frac{\mathbf{r}_k}{r_k^3}. \tag{35}$$

One of the canonical equations spanned by F_k is

$$\dot{\mathbf{r}}_k = \frac{\mathbf{p}_k}{\beta_k} \tag{36}$$

This equation apparently contradicts the statements done after which $\mathbf{r}_k$ is the astrocentric radius vector and $\mathbf{p}_k$ is the barycentric linear momentum. It only means that the variation of $\mathbf{r}_k$, in the reference Keplerian motion defined by the Hamiltonian (34) is not the actual relative velocity of the kth body, but $\mathbf{p}_k/\beta_k$. This means that, at variance with other formulations, the Keplerian motions defined by Eq. (34) are not tangent to the actual motions.

7   Action-angle variables: Delaunay elements

In both cases, the 'undisturbed' solution is a set of N Keplerian motions. In the case of the two reductions shown in the previous section, the generic Hamiltonian is

$$F = \frac{1}{2} \frac{\mathbf{p}^2}{\beta} - \frac{\mu \beta}{r} \tag{37}$$

where the subscripts have been omitted for the sake of simplicity. The purpose of this and the forthcoming sections is to obtain the Delaunay variables and the Keplerian elements which correspond to the systems of canonical variables introduced before, which must be used when a canonical perturbation theory is constructed using H_0 as the unperturbed approximation. For that purpose, we have to solve the corresponding Hamilton-Jacobi equation and construct the action-angle variables of the given problem. We only give the steps more important to stress the variables appearing in the definitions of the action-angle variables and in the associated Delaunay elements. To that end, the study of the planar case is sufficient and preferable since the rotations necessary when the spatial case is considered, although trivial, introduce many new equations. All conceptual questions appear in the planar case and have the advantage of making the calculations much easier, allowing the crucial points to be clearly identified. The extension to three dimensions can be later done by introducing suitable rotations of the axes. However, once the conceptual problems are solved in the planar case, the usual three dimensional equations can easily be adapted to give the remaining action-angle variables

The given Hamiltonian is separable in polar coordinates. In that case, the part corresponding to the kinetic energy in H_0 is

$$T = \frac{\beta}{2}(\dot{r}^2 + r^2\dot{\psi}^2) \tag{38}$$

where $\dot{r}, \dot{\psi}$ are the variations of the coordinates r, ψ in the corresponding Keplerian motion. If the momenta $p_r = \partial T/\partial \dot{r}, p_\psi = \partial T/\partial \dot{\psi}$ are introduced, we obtain

$$T = \frac{1}{2\beta}\left(p_r^2 + \frac{p_\psi^2}{r^2}\right). \tag{39}$$

The potential energy term is given by

$$U(r) = -\frac{\mu\beta}{r} \tag{40}$$

and the resulting Hamilton-Jacobi equation is the classical one of the two-body problem with β instead of m and μ instead of $G(M+m)$:

$$H = \frac{1}{2\beta}\left(p_r^2 + \frac{p_\psi^2}{r^2}\right) - \frac{\mu\beta}{r}. \tag{41}$$

The solution of this equation is well known. This equation is separable into:

$$p_r = \sqrt{2\beta(E + \frac{\mu\beta}{r}) - \frac{C^2}{r^2}} \tag{42}$$

$$p_\psi = C \tag{43}$$

(for details, see Ferraz-Mello, 2007, Chap. 3). $E = H$ and $C = \mathbf{r} \times \mathbf{p}$ are integration constants that generalize to this case the constant energy and angular momentum of the two-body problem.

The actions associated with the given Hamiltonian are

$$J_r = \frac{1}{2\pi} \oint p_r dr \qquad\qquad J_\psi = \frac{1}{2\pi} \oint p_\psi d\psi. \qquad (44)$$

The integral giving the radial action J_r may be easily calculated if we introduce the constants

$$a \stackrel{\text{def}}{=} -\frac{\mu\beta}{2E}, \qquad (45)$$

$$e \stackrel{\text{def}}{=} \sqrt{1 + \frac{2EC^2}{\mu^2\beta^3}}, \qquad (46)$$

which are, respectively, the mean distance (or *semi-major axis*) and *eccentricity*, and the angle u (*eccentric anomaly*) defined through

$$r = a(1 - e\cos u). \qquad (47)$$

An elementary calculation gives

$$p_r = \sqrt{-2\beta E}\,\frac{e\sin u}{1 - e\cos u}$$

and

$$J_r = \frac{1}{2\pi}ae^2\sqrt{-2\beta E}\int_0^{2\pi} \frac{\sin^2 u\,du}{1 - e\cos u},$$

whose solution is trivial. We thus obtain

$$J_r = -C + \mu\beta\sqrt{\frac{\beta}{-2E}} \qquad\qquad J_\psi = C. \qquad (48)$$

The resulting Delaunay actions are:

$$\begin{aligned} L &= J_r + J_\psi = \beta\sqrt{\mu a} \\ G &= J_\psi \qquad = \beta\sqrt{\mu a}\sqrt{1 - e^2}. \end{aligned} \qquad (49)$$

Since, in general, the planets do not move in the same plane, we have to introduce the inclinations I of their planes of motion over a common reference plane and add the third Delaunay action

$$H_{(\text{Del})} = \beta\sqrt{\mu a}\sqrt{1 - e^2}\cos I. \qquad (50)$$

The Delaunay angles $\ell, \omega = \psi - v$ (and Ω) are obtained in the usual way.

8 Keplerian elements

For each planet, we may transform $\beta, \mu, \mathbf{r}, \mathbf{p}$ into the Keplerian elements a, e, λ, ϖ using the same transformations used to define the ordinary osculating[3] astrocentric elements $a_{\text{osc}}, e_{\text{osc}}, \lambda_{\text{osc}}, \omega_{\text{osc}}$ of the two-body problem as functions of $m, G(M+m), \mathbf{r}, m\dot{\mathbf{r}}$. However, the equations giving the osculating astrocentric elements depend on m only through μ. In order to use always the same routines, the above equations may be transformed. We replace E and C, in Eqs. (45) and (46), by their definitions $E = F$ and $C = \mathbf{r} \times \mathbf{p}$. We obtain

$$a = \frac{r\mu}{(2\mu - rw^2)} \tag{51}$$

$$e = \sqrt{\left(1 - \frac{r}{a}\right)^2 + \frac{(\mathbf{r}.\mathbf{w})^2}{\mu a}} \tag{52}$$

where we have used the velocity in the reference Keplerian motion

$$\mathbf{w} = \frac{\mathbf{p}}{\beta} \tag{53}$$

instead of the actual planetary velocity[4].

The angles are obtained with the usual equations. In the planar problem, the true longitude is given by the angle formed by the radius vector and the first axis of the reference system (to be obtained through $\arctan y/x$ where x, y are the components of $\mathbf{r}$). The anomalies are also easily obtained, starting with the eccentric anomaly (u), which is given by

$$u = \arctan\left(\sqrt{\frac{a}{\mu}} \frac{\mathbf{r}.\mathbf{w}}{a - r}\right). \tag{54}$$

The true (v) and mean (ℓ) anomalies are obtained by means of classical 2-body equations. The other angles to determine are the longitude of pericenter (ϖ) and the mean longitude (λ).

The elements of the reference Keplerian orbit at the time t are a, e, ϖ, λ. Since the parameter λ is variable, it is convenient to substitute it by the so-called 'mean longitude at the epoch' (λ_0), which is the value of λ at a standard 'epoch' t_0.

$$\lambda = \lambda_0 + n(t - t_0) \tag{55}$$

---

[3]The word *osculating* means that not only the so-called 'osculating' orbit has tangent to the actual orbit, but also that the motion on both orbits has the same velocity at the tangency point; however, the curvature radii of the two orbits at the tangency point are different and therefore the orbits are not actually osculating curves in the 3D-space.

[4]Important auxiliary equations in the determination of Keplerian elements are the energy integral written as

$$-\frac{\mu}{2a} = \frac{1}{2}\mathbf{w}^2 - \frac{\mu}{r}$$

and

$$-e\sin u = \frac{rw}{\sqrt{ar}}.$$

where n is the mean-motion (see Section 9).

In the spatial problem, some rotations are necessary beforehand to pass from the reference plane to the orbital plane.

8.1 Application to Jacobi's and Poincaré's canonical variables

8.1.1 Jacobi's canonical variables

The first point to be noted in the case of Jacobi's canonical variables, is that the position vectors are the Jacobi coordinates ϱ_k and the velocity vectors are

$$\mathbf{w}_k = \dot{\varrho}_k \qquad (56)$$

(see Eq. 23). This means that Keplerian orbits constructed with Jacobian relative coordinates are osculating (see Footnote 3).

In actual calculations, it is worthwhile to have the Jacobian coordinates ϱ_k and velocities $\mathbf{w}_k$ given as functions of the astrocentric coordinates and velocities. From Eqs. (8) and (6), we obtain

$$\varrho_{k+1} = \varrho_k + [\mathbf{r}_{k+1} - \mathbf{r}_k] - \frac{m_k}{\sigma_k}\varrho_k, \qquad (57)$$

and, by differentiation,

$$\mathbf{w}_{k+1} = \mathbf{w}_k + [\mathbf{v}_{k+1} - \mathbf{v}_k] - \frac{m_k}{\sigma_k}\mathbf{w}_k, \qquad (58)$$

These equations allow the ϱ_k and $\mathbf{w}_k$ to be constructed recurrently starting with $\varrho_1 = \mathbf{r}_1$ and $\mathbf{w}_1 = \mathbf{v}_1$.

The variations of the Keplerian elements obtained using astrocentric or Jacobi coordinates are shown in Figure 6. The fictitious system used as an example, with planet masses enhanced to magnify the perturbation effects, is based on the actual system GJ 876 (same star mass and initial semi-major axes, but initialy circular astrocentric orbits). The figure shows that the Jacobi elements are more stable than astrocentric elements, and are thus a better choice in Keplerian fits. Since Keplerian fits do not take perturbations into account, they are not linked to a given epoch and the results of simulations completed using astrocentric elements may depend critically on the epoch chosen (which is arbitrary). However, one may keep in mind that, notwithstanding the fact that the bulk formula used in Keplerian fitting is the same in both cases, the transformation of the amplitude K into masses is not given by the same equations in both cases (see e.g. Lee and Peale, 2003; Beaugé et al. 2007)

8.1.2 Poincaré's canonical variables

In the case of Poincaré's relative canonical coordinates, the Keplerian motion corresponding to Eq. (34) is obtained with the ordinary routines substituting

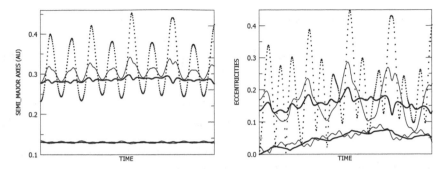

Figure 6. *Semi-major axes and eccentricities of the 2 companions of a system identical to GJ 876 but with planet masses enhanced ($m_1 = 20 M_{\mathrm{Jup}}$ and $m_2 = 5 M_{\mathrm{Jup}}$) to magnify the perturbation effects. Thick line: Keplerian elements constructed using Jacobi variables. Thin line: Idem with Poincaré variables. Dots: Ordinary osculating astrocentric elements. Top lines: Outer planet. Bottom lines: Inner planet. The astrocentric elements of the inner planet coincide with those constructed with Jacobi variables. The total time span shown is 1 year.*

the astrocentric velocities by

$$\mathbf{w}_k = \frac{m_k}{\beta_k} \mathbf{V}_k \tag{59}$$

where $\mathbf{V}_k$ is the absolute (i.e. barycentric) velocity. The position vectors are the relative astrocentric vectors $\mathbf{r}_k$.

The variations of the Keplerian semi-major axes and eccentricities obtained using Poincaré coordinates are also shown in Figure 6.

9 Kepler's third law

The transformation of the previous sections allows the function H to be written as a function of the Keplerian elements and the use of classical expansions. One should, however, always bear in mind that these elements differ from the astrocentric osculating elements by terms of the order of the planetary masses, and that these differences should be taken into account when comparing analytical results with elements obtained from observations (which are often given without any indication of what they are) or numerical simulations.

From the equations obtained for the actions and the Delaunay elements, we obtain the Hamiltonian in terms of the new actions:

$$E = -\frac{\mu^2 \beta^3}{2(J_r + J_\psi)^2} = -\frac{\mu^2 \beta^3}{2L^2} \tag{60}$$

and then, the mean-motion,

$$n = \dot{\ell} = \frac{\mu^2 \beta^3}{L^3} = \sqrt{\frac{\mu}{a^3}}.$$ (61)

The third Kepler's law is thus

$$n^2 a^3 = \mu.$$ (62)

It is important to keep in mind that the mean motions thus defined are not the same if we use Poincaré or Jacobi variables.

10   Conservation of angular momentum

If the only forces acting on the N+1 bodies are their point-mass gravitational attractions, the angular momentum

$$\mathcal{L} = \sum_{k=0}^{N} m_k \mathbf{X}_k \times \dot{\mathbf{X}}_k$$ (63)

is conserved. In a barycentric frame, $\sum_0^N m_k \mathbf{X}_k = \sum_0^N m_k \dot{\mathbf{X}}_k = 0$, and the above equation gives

$$\mathcal{L} = \sum_{k=1}^{N} \mathbf{r}_k \times \mathbf{p}_k,$$ (64)

that is

$$\mathcal{L} = \sum_{k=1}^{N} \beta_k \sqrt{\mu_k a_k (1 - e_k^2)}\, \mathbf{K}_k$$ (65)

where $\mathbf{K}_k$ are the unit vectors normal to the Keplerian planes. This is an exact conservation law and a_k and e_k are the Keplerian elements defined by equations (51) – (52) when $\mathbf{w}_k$ are the absolute velocities corrected by the factor m_k/β_k (see Eq. 59).

The conservation law given by Eq. (64) is also true if Jacobian coordinates are used. In that case, the Keplerian elements are defined by Eqs. (51) – (52) but $\mu_k = G\sigma_k$ and $\mathbf{w}_k = \dot{\boldsymbol{\varrho}}_k$. Indeed, Eqs. (10) and (13) give

$$\mathcal{L} = \sum_{k=1}^{N} \boldsymbol{\varrho}_k \times \boldsymbol{\pi}_k + \frac{1}{\sigma_N} m_k \boldsymbol{\varrho}_k \times \boldsymbol{\pi}_0.$$ (66)

But, Eqs. (13) – (14) also show that

$$\boldsymbol{\pi}_0 = \sum_{k=0}^{N} \mathbf{p}_k = 0.$$ (67)

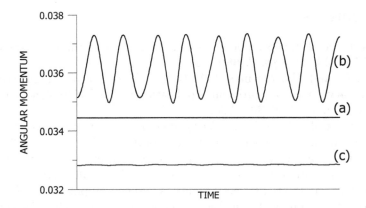

Figure 7. *(a) Constant Angular Momentum $\mathcal{L}$ of the same coplanar system considered in Figure 6. (b) Value of $\widehat{\mathcal{L}}$ as given by Eq. (68) using astrocentric orbital elements. (c) Value of $\mathcal{L}$ which results when the Kepleriam elements are obtained using in Eqs. (51) – (52) but replacing $\mathbf{w}_k$ by the barycentric velocities $\mathbf{V}_k$. The results obtained using Eq. (65) in the cases of both Jacobi and Poincaré elements coincide with (a).*

However, the expression

$$\widehat{\mathcal{L}} = \sum_{k=1}^{N} m_k \sqrt{\mu_k a_k (1 - e_k^2)}\, \mathbf{K}_k \tag{68}$$

where a_k and e_k are the astrocentric osculating elements (Keplerian elements defined by Eqs. (51) – (52) when $\mu_k = G(m_0 + m_k)$ and $\mathbf{w}$ are the astrocentric velocities), often found in the literature, is not an exact conservation law. One may easily see that:

$$\widehat{\mathcal{L}} = \mathcal{L} - \sum_{i=1}^{N} m_i \mathbf{X}_0 \times \dot{\mathbf{X}}_0 \tag{69}$$

showing that the quantity $\widehat{\mathcal{L}}$ differs from the exact angular momentum of the system by a quantity of order $\mathcal{O}(m_i^2)$. Figure 7 shows the values we obtain for $\mathcal{L}$ and $\widehat{\mathcal{L}}$.

Acknowledgments

SFM thanks Bonnie Steves, Martin Hendry and Andrew Collier Cameron for their invitation to participate in this very successful and stimulating summer school. We acknowledge with thanks the chapter improvements suggested by G.G. da Silva.

References

Barnes, R. et al. (2008), The successful prediction of the extra-solar planet HD 74156d, *Astroph. J.* **680**, L57.

Beaugé, C. and Michtchenko, T.A. (2003), Modelling the high-eccentricity planetary three-body problem. Application to the GJ876 planetary system, *Mon. Not. R. A. S.* **341**, 760.

Beaugé, C. et al. (2003), Extra-solar planets in mean-motion resonance: apses alignment and asymmetric stationary solutions, *Astroph. J.* **593**, 1124.

Beaugé, C. et al. (2006), Planetary migration and extra-solar planets in the 2/1 mean-motion resonance, *Mon. Not. R. A. S.* **365**, 1160.

Beaugé, C. et al. (2007), Planetary masses and orbital parameters from radial velocity measurements, In *Extra-solar Planets: Formation, Detection and Dynamics*, (R. Dvorak, ed.) (Wiley-VCH, Weinheim), 1.

Beaugé, C. et al. (2008), Reliability of orbital fits for resonant extra-solar planetary systems: the case of HD82943, *Mon. Not. R. A. S.* **385**, 2151.

Butler, R.P. et al. (2006), Catalog of nearby exoplanets, *Astroph. J.* **646**, 505.

Butler, R.P. et al. (2008) http://exoplanets.org/planets.shtml (2008, Jan. 26th update).

Callegari Jr. N. et al. (2006), Dynamics of two planets in the 3/2 mean-motion resonance: Application to the planetary system of the pulsar PSR B1257+12, *Celest. Mech. Dyn. Astron.* **94**, 381.

Cochran, W.D. et al. (2007), A planetary system around HD 155358: the lowest metallicity planet host star, *Astroph. J.* **665**, 1407.

Dvorak, R. (ed.) (2007), *Extra-solar Planets: Formation, Detection and Dynamics*, (Wiley-VCH, Weinheim).

Ferraz-Mello, S. and Michtchenko, T.A. (2003), *Rev. Mexic. Astron. Astroph.* (ser.conf.) **14**, 7.

Ferraz-Mello, S. et al. (2003), Evolution of migrating planet pairs in resonance, *Celest. Mech. Dyn. Astron.* **87**, 99.

Ferraz-Mello, S. et al. (2005a), The orbits of the extra-solar planets HD 82943c and b, *Astroph. J.* **621**, 473.

Ferraz-Mello, S. et al. (2005b), Chaos and stability in planetary systems, In *Lecture Notes in Physics*, **683**, 219.

Ferraz-Mello, S. et al. (2006), Regular motions in extra-solar planetary systems, In *Chaotic Worlds: From Order to Disorder in Gravitational N-Body Systems* (B.A. Steves and A. Maciejewicz, eds.), (Springer, Dordrecht) 255.

Ferraz-Mello, S. (2007) *Canonical Perturbation Theories: Degenerate Systems and Resonances* (Springer, New York).

Ferraz-Mello, S. et al.(2008), Tidal friction in close-in satellites and exoplanets: The Darwin theory re-visited, *Celest. Mech. Dyn. Astron.* **101**, 171.

Fischer, D. et al. (2008), Five planets orbiting 55 Cancri, *Astroph. J.* **675**, 790.

Gaudi, B.S. et al. (2008), Discovery of a Jupiter/Saturn analog with gravitational microlensing, *Science* **319**, 927.

Greenberg, R. and Brahic, A.(eds.) (1984) *Planetary Rings*, (U. Arizona Press, Tucson)

Greenberg, R. (1987), Galilean satellites - evolutionary paths in deep resonance, *Icarus*, **70**, 334.

Gregory, P.C. (2007), A Bayesian periodogram finds evidence for three planets in HD 11964, *Mon. Not. R. A. S.* **381**, 1607.

Goździewski, K. and Konacki, M. (2006), Trojan pairs in the HD 128311 and HD 82943 planetary systems?, *Astroph. J.* **647**, 473.

Hadjidemetriou, J.D. (2002), Resonant periodic motion and the stability of extra-solar planetary systems, *Celest. Mech. Dyn. Astron.* **83**, 141.

Hadjidemetriou, J.D. (2006), Symmetric and asymmetric librations in extra-solar planetary systems: a global view, *Celest. Mech. Dyn. Astron.* **95**, 225.

Hadjidemetriou, J.D. (2008), On periodic orbits and resonance in extra-solar planetary systems, *Celest. Mech. Dyn. Astron.* **102**, 69

Hagihara, Y. (1970), *Celestial Mechanics*, (MIT Press, Cambridge), vol. I.

Hori, G.-I. (1985), Mutual perturbations of 1:1 commensurable small bodies with the use of the canonical relative coordinates, In *Resonances in the Motion of Planets, Satellites and Asteroids* (S.Ferraz-Mello & W.Sessin, eds.), (IAG-USP, São Paulo), 53.

Kley, W. (2003), Dynamical evolution of planets in disks, *Celest. Mech. Dyn. Astron.* **87**, 85.

Konacki, M. and Wolszczan, A. (2003), Masses and orbital inclinations of planets in the PSR B1257+12 system, *Astroph. J.* **591**, L147.

Laskar, J. (1990), Systèmes de variables et éléments, In *Les Méthodes Modernes de la Mécanique Céleste* (D.Benest & Cl.Froeschlé, eds), Ed. Frontières, 63.

Laskar, J. (2000), On the spacing of planetary systems, *Phys. Rev. Let.* **84**, 3240.

Laughlin, G. et al. (2005), The GJ 876 planetary system: a progress report, *Astroph. J.* **622**, 1182.

Lee, M.H. and Peale, S. (2002), Dynamics and origin of the 2:1 orbital resonances of the GJ 876 planets, *Astroph. J.* **567**, 596.

Lee, M.H. and Peale, S. (2003), Secular evolution of hierarchical planetary systems, *Astroph. J.* **592**, 1201.

Lee, M.H. et al. (2006), On the 2:1 orbital resonance in the HD 82943 planetary system, *Astroph. J.* **641**, 1178.

Lovis, C. et al. (2006), An extra-solar planetary system with three Neptune-mass planets, *Astron. Astrophys.* **441**, 305.

Mayor, M. et al. (2004), The CORALIE survey for southern extra-solar planets. XII. Orbital solutions for 16 extra-solar planets discovered with CORALIE, *Astron. Astrophys.* **415**, 391.

Mayor, M. et al. (2008), The HARPS search for southern extra-solar planets.

XIII. A planetary system with 3 super-Earths (4.2, 6.9, and 9.2 M_{Earth}) *Astron. Astrophys.* **493**, 639.

Michtchenko, T.A. and Ferraz-Mello, S. (2001), Resonant structure of the outer solar system in the neighborhood of the planets, *Astron. J.* **122**, 474.

Michtchenko, T.A. and Malhotra, R. (2004), Secular dynamics of the three-body problem: application to the υ Andromedae planetary system, *Icarus* **168**, 237.

Michtchenko, T.A. et al. (2006a), Modeling the 3-D secular planetary three-body problem. Discussion on the outer υ Andromedae planetary system, *Icarus* **181**, 555.

Michtchenko, T.A. et al. (2006b), Stationary orbits in resonant extra-solar planetary systems, *Celest. Mech. Dyn. Astron.* **94**, 411.

Michtchenko, T.A. et al. (2007), Dynamics of the extra-solar planetary systems, In *Extra-solar Planets. Formation, Detection and Dynamics*, (R. Dvorak, ed.) (Wiley-VCH, Weinheim), 151.

Michtchenko, T.A. et al. (2008a), Dynamic portrait of the planetary 2/1 mean-motion resonance - I. Systems with a more massive outer planet, *Mon. Not. R. A. S.* **387**, 747.

Michtchenko, T.A. et al. (2008b), Dynamic portrait of the planetary 2/1 mean-motion resonance - II. Systems with a more massive inner planet, *Mon. Not. R. A. S.* **391**, 215.

Niedzielski, et al. (2008) *Astrophys. J.Letters* (submitted)

Pauwels, T. (1983), Secular orbit-orbit resonance between two satellites with non-zero masses, *Celest. Mech.* **30**, 229.

Poincaré, H. (1905), *Leçons de Mécanique Céleste*, (Gauthier-Villars, Paris), vol.I

Psychoyos, D. and Hadjidemetriou, J.D. (2004), Dynamics of extra-solar systems at the 5/2 resonance: application to 47 UMa, *Proc. IAU Symposium* **197**, 55.

Psychoyos, D. and Hadjidemetriou, J.D. (2005), Dynamics of 2/1 resonant extra-solar systems application to HD 82943 and GJ 876, *Celest. Mech. Dyn. Astron.* **92**, 135.

Rivera, E.D. et al. (2005), A 7.5M_{Earth} Planet orbiting the nearby star, GJ 876, *Astroph. J.* **634**, 625.

Short, D. et al. (2008), New solutions for the planetary dynamics in HD160691 using a Newtonian model and latest data, *Mon. Not. R. A. S.* **386**, L43.

Tinney, C.G. et al. (2006), The 2:1 resonant exoplanetary system orbiting HD 73526, *Astroph. J.* **647**, 594.

Udry, S. et al. (2007), The HARPS search for southern extra-solar planets. XI. Super-Earths (5 and 8 M_{Earth}) in a 3-planet system, *Astron. Astrophys.* **469**, L43.

Vogt, S.S. et al. (2005), Five new multicomponent planetary systems, *Astroph. J.* **632**, 638.

Voyatzis, G. and Hadjidemetriou, J.D. (2005), Symmetric and asymmetric librations in planetary and satellite systems at the 2/1 resonance, *Celest. Mech. Dyn. Astron.* **93**, 263.

Voyatzis, G. and Hadjidemetriou, J.D. (2006), Symmetric and asymmetric 3:1 resonant periodic orbits with an application to the 55 Cnc extra-solar system, *Celest. Mech. Dyn. Astron.* **95**, 259.

Wittenmyer, R.A. et al. (2005), Long-period objects in the extra-solar planetary systems 47 Ursae Majoris and 14 Herculis, *Astroph. J.* **654**, 625.

Wright, J.T. et al. (2007), Four new exoplanets and hints of additional substellar companions to exoplanet host stars, *Astroph. J.* **657**, 533.

Yuasa, M. and Hori,G.-I. (1979), New approach to the planetary theory, *IAU Symp.* **81**, 69.

Zhou, L.Y. et al. (2004), Apsidal corotation in mean motion resonance: the 55 Cancri system as an example, *Mon. Not. R. A. S.* **350**, 1495.

Zhou, L.Y. et al. (2008), Formation and transformation of the 3:1 mean-motion resonance in 55 Cancri System, *Proc. IAU Symposium* **249**, 485.

The stability of terrestrial planets in planetary systems

Rudolf Dvorak, E. Pilat-Lohinger, R. Schwarz and Ch. Lhotka

ADG, Department of Astronomy, University of Vienna, Austria

1 Introduction

One main topic in astronomy since the first discovery of an extra-solar planet near a sun-like star in 1995 is the formation and dynamical evolution of planets. Although up to now we have not detected any planet near the size of our Earth, numerous dynamical studies have been undertaken to search for the dynamically stable orbits in systems with large planets. Special emphasis here is orientated on the stability zones of celestial bodies moving in the so-called habitable zone (=HZ)[1] around a star. Such studies are important, especially with regard to present (CoRoT) and forthcoming (like e.g. Kepler, Darwin, TPF) space missions that try to find a planet – similar to our Earth – where complex life could have developed. Since we know that the development of a biosphere and the further evolution to intelligent life-form is a process on a long time-scale, the long-term stability of such a planet is an important and necessary condition for the habitability of a planet.

In this lecture we first introduce the dynamical models. Then the different methods of our investigations for stable orbits of fictitious terrestrial planets will be introduced and finally we discuss in brief the results.

For the dynamics of the systems, which sometimes have planets with large eccentric orbits, we refer to Chapter 12 by Ferraz-Mello et al. in this volume.

2 The dynamical models

The majority of planetary systems we know – up to now – host at least one large planet comparable to Jupiter or even larger. For the study of additional

---

[1] Where water may exist in liquid form on the surface.

small planets in these systems we therefore use the elliptic restricted three-body problem (ER3BP) as the basic model, to investigate their dynamical behavior. It is defined as follows: A massless regarded body m_3 moves in the gravitational field of two massive bodies (m_1, m_2); these bodies, called primaries are moving on Keplerian orbits around their common barycentre. If one of their orbits is circular we speak of the circular restricted three body problem and if the motion of the third body is not in the plane of motion of the primaries one speaks of the spatial restricted three-body problem.

The equations of motion for m_3 in the circular and planar case are given in a rotating coordinate system (Figure 1) with respect to another inertial $x - y$ coordinate system. Here the primaries have fixed positions on the ξ-axis via:

$$\ddot{\xi} - 2n\dot{\eta} = \frac{\partial \Omega}{\partial \xi} \ and \ \ddot{\eta} + 2n\dot{\xi} = \frac{\partial \Omega}{\partial \eta} \tag{1}$$

where Ω is the potential, (ξ, η) is the position vector of the massless body and n means the constant mean motion of one primary around the other. The problem of finding the orbit for the third massless body m_3 is to find solutions $\xi(t)$ and $\eta(t)$ for this system of second order differential equations. Fortunately it is possible to find an integral of motion in this case. We multiply the first of the two equations with $\dot{\xi}$, the second one with $\dot{\eta}$ and add these two new equations where the term $2n\dot{\eta}\dot{\xi}$ cancels:

$$\ddot{\xi}\dot{\xi} + \ddot{\eta}\dot{\eta} = \frac{\partial \Omega}{\partial \xi}\dot{\xi} + \frac{\partial \Omega}{\partial \eta}\dot{\eta} \tag{2}$$

This equation can be integrated, because the right hand side is the total derivative with respect to the time and leads to

$$2\dot{\xi}^2 + \dot{\eta}^2 \equiv v^2 = 2\Omega - C \tag{3}$$

where $C(\xi, \eta, \dot{\xi}, \dot{\eta})$ is the Jacobian constant[2]. With the aid of the integral of motion C, the dimension of the system may be reduced by one degree. Although this system poses the former mentioned integral of motion which 'replaces' the energy, no closed solution for the motions can be derived, in general.

We should mention the following quite interesting property of the restricted three body problem. If – in addition to the Jacobian constant – one could derive a second integral of motion, then the problem would be integrable! Let us denote it by $D(\xi, \eta, \dot{\xi}, \dot{\eta}) = const$. We could then express the velocities by

$$\frac{d\xi}{dt} = f(\xi, \eta) \ and \ \frac{d\eta}{dt} = g(\xi, \eta). \tag{4}$$

By division one differential equation remains, namely

---

[2]After Jacobi (1804–1851).

$$\frac{d\eta}{d\xi} = \frac{g}{f} = h(\xi, \eta) \tag{5}$$

which then could be solved with the aid of a factor of the form $M(\xi, \eta)$ such that $0 = fMd\eta - gMd\xi = dF$, would be a complete differential expression from this it follows that $F(\xi, \eta) = const$ which is the equation of the orbit we are looking for. With the aid of expression (5) one could eliminate from (4) the coordinates

$$\dot{\xi} = \phi(\xi), \ \dot{\eta} = \psi(\eta) \tag{6}$$

and from this one easily via a quadrature could find the final solution $\xi = \xi(t)$ and $\eta = \eta(t)$. Since in our case no additional integral of motion is known, this way of solution cannot be used. One should also mention that – even with an additional integral of motion in a two degrees of freedom system it may be impossible to solve Eq. (5), because the factor M is rather difficult, if not impossible to find.

We thus refer to the method of numerical integration of the equations of motion for a great variety of initial conditions, to be able to have a detailed picture of the respective phase space. For example, we know that mean motion resonances are acting quite strong in the asteroid belt, they form the so-called Kirkwood gaps (Figure 2). Numerical integrations can reproduce and explain this curious behavior.

Analytical results can be derived for motions in resonances via special techniques (like used for the 2:1, the 3:1 and the 3:2 resonance; e.g. Ferraz-Mello (1999)) where one can determine the region of stable motion inside such a resonance. Extensive work exists for the asteroids in the main belt, but no study for terrestrial planets (TPs) in extra-solar planetary systems (EPSs) have been presented so far.

A quite well studied resonance is the 1:1 resonance (e.g. Érdi (1988), Érdi (1996) Marchal (1990)), where one can find stable equilibrium points (L_4 and L_5), which form an equilateral triangle with the primaries (Figure 3). In the vicinity of these points we have two 'clouds' of bodies always moving 60° ahead or 60° behind Jupiter in its orbit. We emphasize that the 3 points (L_1, L_2 and L_3) on the connecting line between the primary bodies are not stable and the points L_4 and L_5 are only stable for a mass ratio of the primary bodies[3] $M_{GG}/M_{star} < 1/25$. So we can say that for all planets in our solar system these points are stable, but only Jupiter and – as we know from recent discoveries – also Neptune's equilateral equilibrium points are populated with asteroids near these equilibrium points.

For studies with more than two massive bodies (a star with more than one gas giant (GG) or a binary system with a massive planet) one needs to use numerical methods, which will be shortly discussed in the next section.

[3] M_{GG} mass of a gas giant like Jupiter given in solar masses and M_{star} mass of the central star also given in solar masses.

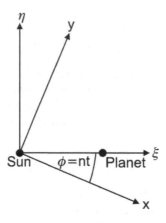

Figure 1. *The restricted three body problem.*

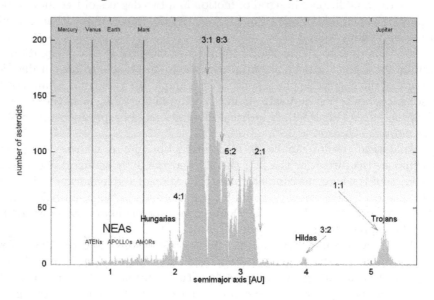

Figure 2. *The distribution of the asteroids in the solar system with respect to the sun and the most important mean motion resonances are marked in this figure.*

3 Numerical methods and analysis of the results

If no analytical results via simplified models apply, the only way for exploring the phase space structure is to accomplish extensive numerical integrations of the respective equations of motion. In the respective numerical simulations widely used are Bulirsch-Stoer and Lie-integration (see Lichtenegger, 1984 or

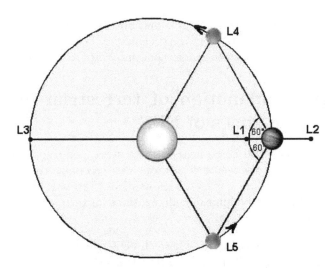

Figure 3. *The 5 equilibrium points of Lagrange (L4,L5 are linearly stable and L1 to L3 are unstable).*

Hanslmeier & Dvorak 1984) methods. There are also free software packages e.g. the 'Mercury6' a hybrid orbit integrator, where a Bulirsch-Stoer integrator is combined with a symplectic one (Chambers 1999). To analyze the results besides a direct check of the orbital elements (especially the eccentricity, which is quite a good parameter in separating stable and unstable orbits for this kind of studies) or the Delaunay elements which present the canonical action angle variables (e.g. used in Schwarz et al. 2004) different chaos indicators can be used to separate between regular and chaotic motion:

1. The Lyapunov Characteristic Exponent (LCE), also called Lyapunov Indicator LCI, e.g. Froeschlé (1984).

2. The Fast Lyapunov Indicator (FLI); Froeschlé et al (1997).

3. The Mean Exponential Growth of Nearby Orbits (MEGNO); Cincotta and Simo (2000).

4. The Smaller Alignment Index SALI; Skokos et al (2004).

5. The Relative Lyapunov Indicator (RLI); Sandor et al (2004).

To get significant stability maps (e.g. Figure 4) for the models mentioned in the previous section ten thousands of fictitious objects have to be integrated simultaneously for a sufficient large grid of initial conditions and a time interval, depending on the method of analysis used. For some chaos indicators some hundred periods of the primaries are sufficient to get a first estimate of the border between regular and chaotic motion, others need much more time.

However, to distinguish between stable and unstable orbits one needs a direct check of the orbital elements (here the eccentricity is the best parameter), but this has to be done for a much longer time interval (10^6 years and more).

4 Regions of motion of terrestrial planets in habitable zones of EPS

From the more than 200 EPSs near sun-like stars (see website of J. Schneider)[4] that have been discovered so far, one can recognize that most of these systems build very simple *single-star single-planet systems*, for which we have to distinguish different dynamical configurations for terrestrial planets (=TP) moving in the HZ:

1. The solar system situation, where the HZ is between the host-star and the GG; we may call this case **inner HZ (IHZ)**.

2. The hot-Jupiter case, where the HZ is outside the orbit of the GG; this will be called **outer HZ (OHZ)**.

3. When GG is moving itself in the HZ we have different possibilities for TPs to be on stable orbits: a TP as a satellite (i.e. habitable Moon) moves in the **satellite HZ (SHZ)** or a TP in a Trojan-like orbit is in the **Trojan HZ (THZ)**.

A mixture of these groups can happen in multi-planetary systems of which we are aware of 25 systems, but we will not treat them in this lecture. The OHZ is less interesting from the dynamical point of view, because we expect only minor perturbations from a hot Jupiter (i.e. a planet orbiting very close and mostly in circular motion about the host-star) on a TP moving in the HZ. The SHZ is also a possible configuration and there exist some dynamical studies using the ER3BP determining an upper limit for a satellite orbit around a planet for possible stable orbits (e.g. Érdi 1988, 1996 and Marchal 1990). In the following we will concentrate on the other two configurations: the IHZ and THZ.

The case of TPs in binaries, where, in addition to a double star, a planet like Jupiter is orbiting one of the primaries, will be treated in one of the following sections. It is evident that for this problem one needs to treat at least a restricted 4-body problem with three massive primaries and fictitious, massless regarded terrestrial planets.

4.1 Terrestrial planets in the IHZ

Many specific systems have already been investigated by different scientists to clarify the question of stable orbits in the HZ (e.g. Menou & Tabachnik

---

[4]http://exoplanet.eu

2003, Érdi 2004, Dvorak 2003a,b, Dvorak et al 2004, Barnes & Raymond 2004, Schwarz et al., 2005, and many others). A quite promising approach is the compilation of a whole catalogue (called Ëxocatalogue) of possible configurations with different mass ratios of the primaries and also different eccentricities of the primaries' orbits. This was undertaken with the aid of extensive numerical integrations in the model of the ER3BP and the results have been published recently (Sandor et al., 2007).

The computation of numerous stability maps (an example is given in Figure 4) allows the application to all single-star single-planet systems and gives a first estimate whether one can expect stable orbits in the HZ. In the respective plot – an initial condition diagram of semi-major axes of the fictitious massless bodies versus the eccentricity of the primaries – the stable motion (characterized by the RLI) is labeled by white and light grey colors, chaotic motion by dark grey to black. The dark vertical stripes in the stable zone correspond to mean motion resonances (MMRs) with the GG – like the ones we know from the astroid belt.

To apply the Ëxocatalogue (available at http://www.univie.ac.at/adg) to a certain EPS one has to determine the mass-ratio of the host-star and the GG to select the appropriate stability map from the catalogue. Taking into account the eccentricity of the giant planet and the HZ of the system (according to the study by Kasting et al., 1993), we get a first view on the stability of the HZ (see the rectangles of the different EPS in Figure 4). A detailed description of the use of the Ëxocatalogue is given in Sandor et al. (2007), where many applications to the known EPS are shown. Of course it can be applied also to every newly discovered single planet in an EPS, where the corresponding stability map indicates if an additional TP could move on a stable orbit. In Figure 4 the units are such, that the Jupiter-like planet has a semi-major axis at 1 [AU]; therefore the habitable zone for the solar system is around 0.2 units.

4.2   Terrestrial planets in the THZ

Theoretical studies predict that Trojans are a common byproduct of planet formation (e.g. Ford & Chiang 2007, Beaugé et al. 2007), and could be possibly formed terrestrial planets in the vicinity of L_4 and L_5. A lot of work has been done to estimate the largeness of the stable region for the Jupiter and the Neptune Trojans and some work has been published recently also for EPS. These kind of orbits have two different periods of libration around their equilibrium points, which depend on the mass ratio of the primaries (see Figure 5).

For terrestrial Trojan planets one can use extensive numerical integrations for a whole grid of initial conditions around the equilibria points L_4 and L_5. We show for different models, depending on the eccentricity of the primaries' orbit and the mass ratio, the extension of the stable regions with respect to the semi-major axis and the angular distance to the Lagrange point itself. Figures 6 and 7 show these stability maps (in a 3 dimensional graph) for mass

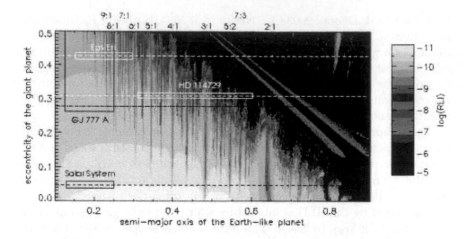

Figure 4. *Stability map out of the Exocatalogue. (Reprinted with permission from Sandor, Z. et al., A stability catalogue of the habitable zones in extrasolar planetary systems, (2007), MNRAS 375, 1495–1502.)*

ratios up to $\mu = 0.04$. Figure 6 presents the 3D graph for an initial eccentricity (Trojan and GG) e=0.0 (left graph) and e=0.05 (like in the case of the Jupiter Trojans). Figure 7 presents higher eccentricities, which could be a problem for the habitability of a terrestrial planet, because for larger eccentricities the planet moves only partly in the HZ. Additionally we compiled a catalogue of stability maps. This catalogue of possible Trojan planets (see also http://www.univie.ac.at/adg/Research/exotro/exotro.html) is useful for any EPSs to be found in the future to give the probability of hosting terrestrial planets around L_4 or L_5.

We also performed analytical investigations, based on Nekhoroshev theory to obtain the nonlinear stability regions around the Lagragian points, which are stable in the sense of Nekhoroshev for a time equal to the age of the solar system. The method based on a symplectic mapping model for the ER3BP in the 1:1 resonance was formulated in a general way and can be applied to solar and extra-solar Trojan configurations (Lhotka et al. 2007). In the respective Figure 8 (top graph) we show the phase portrait of the surface of section in the ER3BP, developed from the generating function based on the averaged Hamiltonian - proposed by Hadjidemetriou (1991). The plots show two sections of the surface of section to cope with the disturbing effects of the 4D-mapping when projected onto 2D space. One clearly sees the well-known chain of islands corresponding to resonant motion of the asteroid as well the librational and rotational behavior (Figure 8 bottom graph). This mapping is based on a 4th (8th) order development of the eccentricities (variation in the semi-major axis), and was iterated in its implicit form to preserve all dynamical behavior of the mapping.

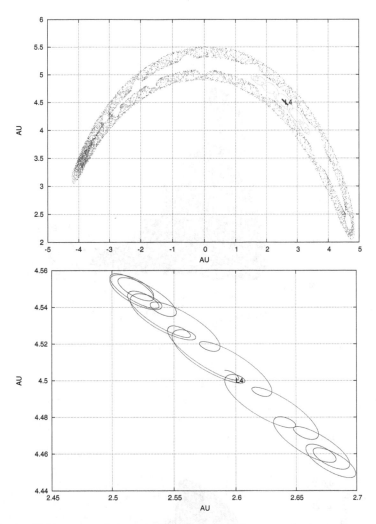

Figure 5. *Orbits around the equilibrium points in a rotating coordinate system*

4.3 Terrestrial planets in double stars

Stability studies of planetary motion in binary systems are very important, since we expect an increasing number of detected planets in such stellar systems in the future – due to the fact that most of the stars in the solar neighborhood form double or multiple star systems. As of October 2007 we know only few planetary systems, which have a 'close' stellar companion, out of a – meanwhile long – list of extra-solar planetary systems (see Raghavan 2006). In most of these double star systems the distance of the 2 stellar components is between 100 and more than 12000 AU; therefore, it is obvious that the detected planets were found to move around one stellar component.

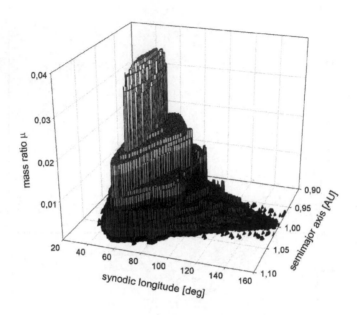

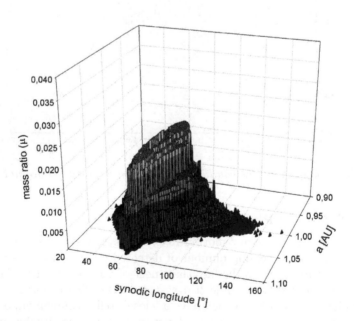

Figure 6. *The stability area for e=0 (top graph) and e=0.05 (Jupiter case, bottom graph).*

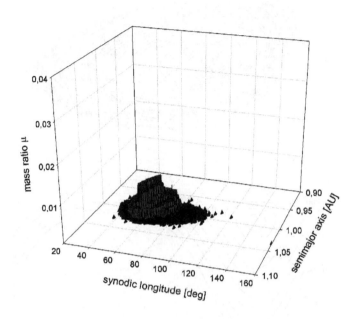

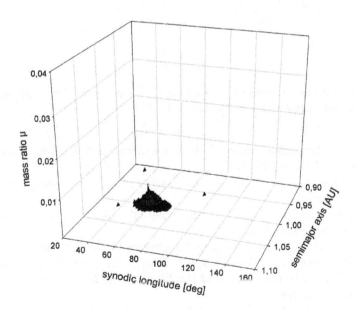

Figure 7. *The stability area for e=0.15 (top graph) and e=0.3 (bottom graph).*

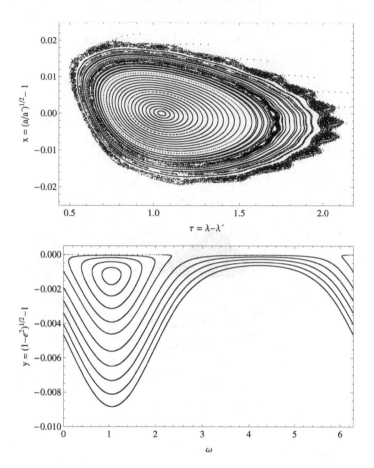

Figure 8. *Two projections of the surface of section of the 4D-mapping around the Lagrangian equilibrium point L_4 or L_5.*

From the dynamical point of view, we distinguish 3 types of motion in such systems (Dvorak, 1984):

(i) the satellite-type (or S-type) motion, where the planet moves around one stellar component;

(ii) the planet-type (or P-type) motion, where the planet surrounds both stars in a very distant orbit; and

(iii) the libration-type (or L-type) motion, where the planet moves in the same orbit as the secondary – locked in 1:1 MMR. Since the stability of L-type motion is limited to mass-ratios of the binary $< 1/25$, this motion is not so interesting for double stars.

Since only planets in S-type motion have been detected so far, we will restrict our discussion to this type and present results of our studies on the three close binaries (γ Cephei, Gl86 and HD41004AB).

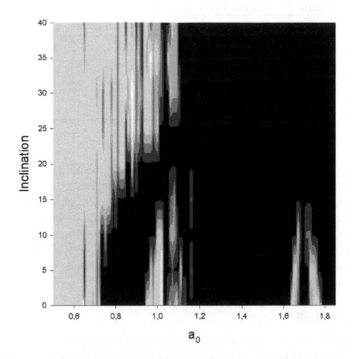

Figure 9. *FLI - stability map of fictitious TPs in the γ Cephei system. The x-axis shows the different semi-major axes of the TPs and the y-axis displays the dependence on the inclination of the TPs. Light grey regions indicate regular motion and the black zone labels the chaotic motion.*

For our numerical investigation of these binary systems we used the restricted 4 body problem (R4BP) – where we study the motion of a massless body in the gravitational field of the primary ($= m_1$), the secondary ($= m_2$) and the giant planet ($= m_3$).

The dynamical state of the orbits was first determined by means of a chaos indicator (the FLI) and then studied through straightforward orbital computations from which we examined the evolution of the orbit's eccentricity and computed the maximum eccentricity over the whole integration time.

4.3.1   γ Cephei

The double star γ Cephei can be found in the constellation Cepheus about 11-12 pc away from our solar system. The detected Jupiter-like planet moves in a S-type orbit around its host-star (a K1 or K2 IV star). The orbital parameters of this system according to the observations by Cochran et al. (2002) are given in Table 1.

Applying the results of a general stability study about planetary motion

Table 1. *Orbital parameters of the binary γ Cephei.*

	primary	secondary	planet
mass [solar masses]:	1.6	0.4	.00168
semi-major axis [AU]:		21.36	2.15
eccentricity:		0.44	0.209
period [years]		70	2.47

in binaries (see e.g. Pilat-Lohinger & Dvorak, 2002) to this system, the border of the stable zone for the parameters given in Table 1 is at about 3.6 AU, so that the discovered planet is well inside the stable region. Since the GG moves in the HZ of γ Cephei[5] in a high eccentric orbit and we know from former studies (see e.g. Dvorak et al., 2003a) that the only dynamically stable region in this system is closer to the host-star and therefore, well inside the orbit of the GG (see the faint regions of Figure 9). This means that, the system allows only habitable moons.

4.3.2   Gliese 86

The binary Gliese 86 is about 11 pc away from the Sun in the constellation Eridanus. The double star system consists of a K1 main sequence star ($m_1 = 0.7 M_S$) and in all probability a white dwarf (with a minimum mass of $0.55 M_S$) at about 21 AU as proposed by Mugrauer & Neuhäuser (2005) using NAOS-CONICA (NACO) and its new Simultaneous Differential Imager (SDI). The planet was found to be very close to the K1 V star, at 0.11 AU with an orbital period of less than 16 days (Queloz et al., 2000). Due to the CORALIE measurements a minimum mass of $4 M_{Jupiter}$ was determined.

As the detected GG moves at 0.11 AU and the HZ is – according to Kasting et al. (1993) – between 0.48 and 0.95 AU in this system, the gravitational influence of the GG on the HZ is not very strong. The most important question for TP in the HZ is, where the GG has been formed[6].

Figure 10 summarizes the results of the maximum eccentricity (max-e) of all TPs in the HZ. One can see constant level lines of max-e for initial eccentricities of the TPs ≥ 0.09. More precisely, the value of the max-e level curve corresponds to the initial eccentricity of the TP ($= e_{planet}$). The region shows the zone of highest max-e values, and the 'finger-like' shape indicates

---

[5]We have to point out that the host-star is already a sub-giant of type K1 or K2 and was previously a late A or early F main sequence star.

[6]If it was formed at a distance between 4 and 5 AU and migrated towards the star through the HZ, an already existing terrestrial-like planet would have been ejected from the system. If the GG was built closer to the star – near the region, where it was found (see Wuchterl et al., 2000), then TPs can be expected in the HZ (but they cannot be detected up to now).

max-e of HZ (WD)

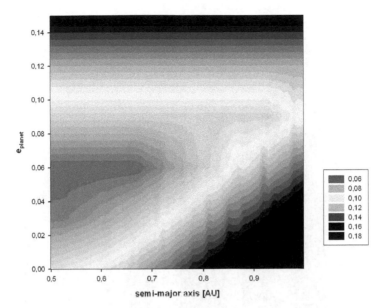

Figure 10. *Maximum eccentricity plot of the HZ of Gliese86 A (x-axis) for fictitious planets with different initial eccentricities (y-axis).*

the region of lowest max-e. The higher values of max-e in the lower right area of Figure 10 result from the very high eccentricity (0.7) of the binary used in our computations.

The border of the so-called 'continuously habitable zone (CHZ)' (i.e. the region, where the whole planetary orbit is in the HZ) depends on the initial eccentricity of the TP in the HZ. Our study has shown that a CHZ exists up to an initial eccentricity of the TPs of 0.33. A more detailed study thereto is in progress.

4.3.3 HD41004 AB

In the visual binary HD41004 AB a planetary companion for HD41004 A was detected in 2003 (Zucker et al. 2003). The giant planet (of 2.3 Jupiter-masses) moves in a highly eccentric orbit ($e_{GG} = 0.39 \pm 0.17$) around the K1V star at a mean distance $a_{GG} \sim 1.31$ AU. The secondary, a M2V dwarf, surrounds the primary in about 20 AU and is accompanied by a close-in brown dwarf with a period of only 1.3 days. For our stability study of TPs moving in the HZ of HD41004 A we calculated for different eccentricities of the binary (0.1 − 0.5) and of the GG (0.22 − 0.39) the dynamical behavior of TPs in the region between 0.45 and 1 AU (= HZ of HD41004 A − according to Kasting et al.,

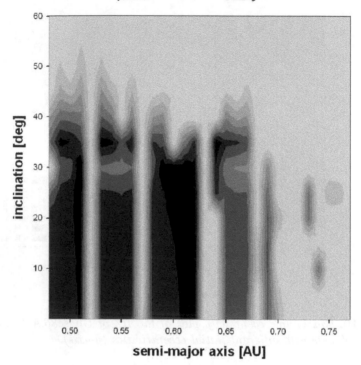

Figure 11. *Maximum eccentricity plots of the HZ of HD41004 A (x-axis) for fictitious planets with different initial inclinations (y-axis). The grey scale indicates the different values of max-e, where the darkest region labels the orbits of lowest max-e.*

1993).

The best case of all stability maps is shown in Figure 11, where the stability region is labeled by the dark areas. One can see that the stable motion is limited to the inner region of the HZ ($a_{TP} < 0.7AU$) and the HZ of HD41004 A is fragmented into several stable stripes. This structure can be explained by MMRs with the detected giant planet – where the 4:1 MMR is near 0.52 AU, the 7:2 MMR is near 0.57 AU, the 3:1 MMR is near 0.63 AU and the 8:3 MMR is near 0.68 AU.

The stability map of Figure 11 represents the result for an eccentricity of the GG of 0.22 (which is the lowest eccentricity of the GG according to the observational data), an eccentricity of the binary of 0.1 and the TPs were started in circular motion. The different grey shades show different values of the max-e of the TPs – the darkest regions label the orbits of lowest max-e values (< 0.2). One can see two regions (around 0.5 and 0.6 AU) where we can expect low-eccentric motion, so that the whole orbit would be in the HZ for at least 10^5 periods. A detailed study of this system can be found in Pilat-Lohinger & Funk (2007).

5 Summary

In this lecture we gave a short overview about the dynamical stability of terrestrial-like planets (TPs) moving in the habitable zone (HZ) of a main-sequence star. Due to the detected extra-solar planets we have to distinguish different HZs from the dynamical point of view. The crucial point is the orbit of the gas giant (GG) of a planetary system; therefore,

(i) an inner HZ (IHZ) describes the solar-system configuration;

(ii) an outer HZ (OHZ) is characteristic for a hot-Jupiter configuration; and

(iii) the satellite and Trojan HZ (SHZ and THZ) characterize the motion of TPs when the GG moves in the HZ.

For the simple *single-star single-planet systems* we concentrated ourselves on two cases, namely TPs in the IHZ and THZ. We presented for both HZ a catalogue of stability maps – in this context we have to note that the Ëxocatalogue (see Section 4.1 and Sandor et al., 2007) contains also stability maps for the OHZ, but they were not discussed in this lecture. Furthermore, we reviewed in brief the planetary motion in the three close binaries: HD41004 AB, Gliese 86 and γ Cephei, where we have to take into account also the perturbations of an accompanying star. We have shown that Gliese 86 is the only system which allows the existence of additional TPs in the HZ without restrictions from the dynamical point of view. From the astrophysical side, we have to consider that the second star is already a white dwarf – and quite close to the host-star – so that its evolution probably has influenced the HZ substantially.

For HD41004 A we could also find conditions for the existence of additional TPs in the HZ, while for γ Cephei we claim that only so-called habitable moons would be possible – but this has still to be studied.

Acknowledgments

RD and EP-L thank the organizers of the school for their hospitality and pleasant ambiance. EP-L, RS and ChL want to acknowledge the support by the Austrian Science Fund (FWF): EP-L for project no. P19569-N16, RS for the Erwin Schrdinger grant J2619-N16 and ChL for grant number P-18930.

References

Barnes, R. et al. (2008) *Astroph. J.* **680**, L57.
Beaugé, C. et al. (2007), *A&A* **463**, 359.

Chambers, J.E. (1999), *MNRAS* **304**, 793.

Cincotta, P.M. & Simo, C., (2000), *Astron.Astrophys.* **147**, 205.

Cochran, W.D. et al (2002), *BAAS* **34**, 42.02.

Dvorak, R. (1984), *CDMA* **34**, 369.

Dvorak, R. et al (2003a), *Astron.Astrophys.* **398**, L1-L4.

Dvorak, R. et al (2003b), *Astron.Astrophys.* **410**, L13-L16.

Dvorak, R. et al (2004), *Astron.Astrophys.* **426**, L37-L40.

Érdi, B. (1988), *Celestial Mechanics* **43**, 303.

Érdi, B. (1996), *Celestial Mechanics* **65**, 149.

Érdi, B. et al (2004), *MNRAS* **351**, 1943.

Ferraz-Mello, S. (1999), *Cel.Mech.Dyn.Astron.* **73**, 25.

Ford, E. B. & Chiang, E. I. (2007), *ApJ* **661**, 602.

Froeschlé, C. (1984), *Celestial Mechanics* **34**, 95.

Froeschlé, C., Lega, E. and Gonczi, R. (1997), *Cel.Mech.Dyn.Astron.* **67**, 41.

Hadjidemetriou, J.D. (1991), in *Predectibility, Stability and Chaos in N-Body Dynamical System* edited by Roy, A.E. Plenum Press, New York, 157.

Hanslmeier, A. & Dvorak, R. (1984), *A&A* **132**, 203.

Kasting, J.F, Whitmire D.P. and Reynolds, R.T. (1993), *ICARUS* **101**, 108.

Lichtenegger, H. (1984), *Cel.Mech.Dyn.Astron.* **34**, 357.

Marchal, C. (1990), *The Three-Body Problem*, Elsevier **49**.

Menou, K. & Tabachnik, S. (2003), *AJ* **583**, 473.

Mugrauer, M. & Neuhäuser, R. (2005), *MNRAS* **361**, L15.

Pilat-Lohinger, E. & Dvorak, R. (2002), *Cel.Mech.Dyn.Astron.* **82**, 143.

Pilat-Lohinger, E. & Funk, B. (2007), *A&A* submitted.

Queloz, D. et al (2000), *A&A* **354**, 99.

Rabl, G. & Dvorak, R. (1988), *A&A* **191**, 385.

Raghavan, D. (2006), *ApJ* **646**, 523.

Sandor, Z. et al (2004), *Cel.Mech.Dyn.Astron.* **90**, 127.

Sandor, Z. et al (2007), *MNRAS* **375**, 1495.

Schwarz, R., Gyergyovits, M. and Dvorak, R. (2004), *CeMDA* **90**, 139.

Schwarz, R. et al. (2005), *Astrobiology Journal* **5**, 579.

Skokos, Ch. et al. (2004), *Journal of Physics A* **37**, 6269.

Wuchterl, G., Guillot, T., and Lissauer, J.J. (2000), in *Protostars and Planets IV* eds. V. Mannings, A.P. Boss and S.S. Russel, Univ. of Arizona Press, 1081.

Zucker, S. et al. (2003), in *A&A* **404**, 775.

Did the two Earth poles move widely 13,000 years ago? An astrodynamical study of the Earth's rotation

Christian Marchal

General Scientific Direction ONERA, BP 72, 92322 Chantillon cedex, France

The last ice age was maximum circa 19,000 BC and its map shows a beautiful symmetry about a "pole" in the present Baffin Sea. The astrodynamical study of the Earth's rotation leads to the concept of a "stable equilibrium position of the poles". The estimation of the Earth inertia matrix during the last ice age is not yet accurate enough to confirm the past position of the pole near the Baffin Sea, but it shows that indeed the pole could have moved that far in a few millennia. This estimation also shows that the Earth polar moment of inertia had increased very slightly during the melting of the last ice age, with a corresponding huge release of energy. This release, with the 120-meter rise of sea level and the upheaval of climate, has contributed to make that age an age of natural disasters.

1   Introduction

The present Earth poles have only very small motions at the Earth's surface: a few meters per year, which is negligible but for specialized observatories. However, they had large displacements in the past and maps of their displacements during geological ages have been drawn. The last ice age, Würm in Eurasia and Wisconsin in North America, is recent. Its maximum occurs circa 19,000 BC, when our Cro-Magnon ancestors were already widely spread on Earth.

The melting of 50 million cubic kilometers of ice occurred essentially between 13,000 BC and 8,000 BC, and the sea level went up 120 meters (see Figure 1).

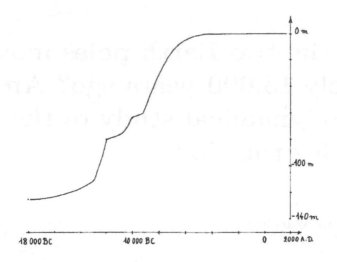

Figure 1. *The rise of average sea level in the past millenia.*

Consider a map of the Würm and Wisconsin glaciations (Figure 2). There is a major dissymmetry. Eastern Siberia was spared, while in America the ice reached New York City at only 40° latitude. It seems that the North Pole was not at its present position, but somewhere near the Baffin Sea.

A symmetrical dissymmetry occurred in the Southern Hemisphere, with much more ice in East Antarctica than in West Antarctica near South America. This would of course imply a recent, large and fast move of the two poles. Is such a move possible? (ten or fifteen degrees!). This is a beautiful astrodynamical question.

This study is essentially based on Marchal (1996) in which an unfortunate misspelling has completely modified the conclusion. We must read **large orogenies and/or large subductions can move the poles by tens** (and not tenths) **of degrees...** This conclusion is applied here to the melting of the quaternary glaciers. The corresponding errata were published in Marchal (1997).

1.1   Preliminary notice (numeration system)

In this chapter with many very large and very small numbers, we will use the convention of numeration by "figures and sizes":

Mass of Earth : M = 5.9737 $\times 10^{24}$ kg = 5.9737 p24 kg. 5.9737 is the figure and p24 the size.

The gravitational constant = 6.6726 $\times 10^{-11} m^3/s^2 kg$ = 6.6726 n11 m^3/s^2 kg. 6.6726 is the figure and n11 the size.

The letter p means "positive power of ten" and the letter n "negative

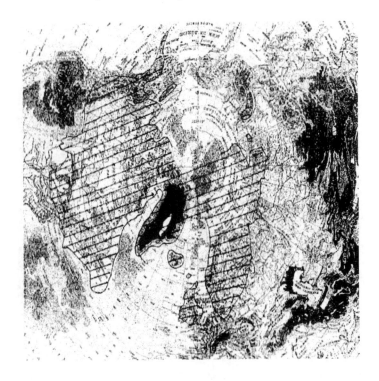

Figure 2. *Polar map of the last ice age maximum (circa 19,000 BC).*

power of ten", **while the figure is always between one and ten** (we cannot write 59 p23 or 0.59 p25 instead of 5.9 p24). The size is thus always well defined. It is of course the most important part of very small and very large numbers. It is even sometimes the only known part.

2   Why this study?

2.1   The geographical extension of the last ice age

Consider the last figure in this chapter, Figure 11, which is the usual presentation of the last maximum of the recent Würm and Wisconsin glaciations (19,000 BC). This Mercator projection of the presentation does not give the essential symmetry of the ice cap. But consider now Figure 2, the same figure in a polar equidistant projection. Of course the ice cap is known only on the continents and the islands. We ignore its true extension on the oceans. Nevertheless the present North Pole is obviously not at the center of the ice cap whose extension is large in Europe (Scandinavia, Northern Russia, Baltikum, British Isles), much more restricted in Asia (Northwestern Siberia) and huge in North America, down to the present position of the Great Lakes and New

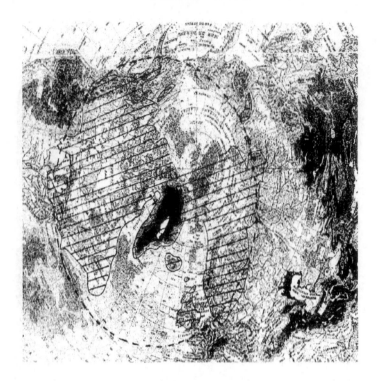

Figure 3. *An idea of the position of the North Pole during the last ice age (near the cross in the Baffin Sea).*

York City at only 40° latitude.

Figure 3 gives an idea of a likely position of the North Pole during the Würm and Wisconsin glaciations, near the cross drawn in the Baffin Sea. The corresponding dotted circle follows very nearly the land limits of the ice cap and the outer glaciers are on mountainous ranges: the Alps, the Verkhoiansk range and the Aleutian range.

Thus two questions follow very naturally: Is it possible that the North Pole was in the vicinity of the Baffin Sea during the last ice age? Is such a fast motion of the pole possible?

2.2   The concept of a stable equilibrium position of the poles

The Earth rotation is stable and the poles have presently only very small motions: they have remained in a domain of 15 meters for more than one century (Figures 4 and 5). Because of the equatorial bulge, the Earth axis is the main axis of largest inertia and this property is accurately verified through the analysis of the Earth gravitational potential (Balmino et al. (1978), Lerch

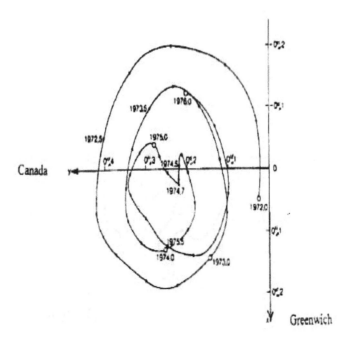

Figure 4. *The motion of the North Pole during the four years 1972-1975. O is a conventional origin. O_x is in the direction of Greenwich and O_y is in the direction of Canada. At the Earth's surface 0.1" is only 3.1 meters.*

et al. (1979), Marsh et al. (1990) and Reigber et al. (1993)).

For rigid bodies, such a rotation is indeed a classical stable rotation; however, the Earth is not a rigid body and its equatorial bulge is a consequence of its rotation.

The full analysis of the stability of the Earth's rotation has been completed in Marchal (1996), with the following results.

For a planet like the Earth, we must consider four types of equilibria (Figure 6):

(A) The hydrostatic equilibrium of a non-rotating planet: There is spherical symmetry; the different layers are spherical and concentric. The density is increasing with depth, or at least it is non-decreasing.

(B) The hydrostatic equilibrium of a slowly rotating planet (Marchal, 1968). If a planet, like the Earth, has a diameter smaller than the distance of synchronous satellites, it is a slowly rotating planet (which is not the case of planet Saturn).

Slowly rotating planets have a hydrostatic equilibrium with "spheroidal

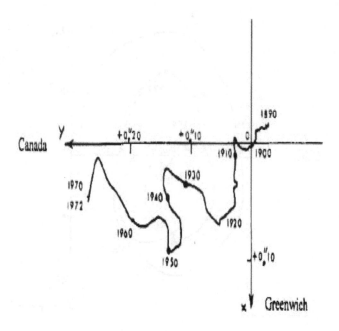

Figure 5. *The slow drift of the North Pole towards Canada when the short period components appearing in Figure 4 are removed.*

symmetry", which is both a symmetry of revolution about the rotation axis and a symmetry about the equatorial plane. The level surfaces are concentric and coaxial oblate quasi-ellipsoid slightly depressed at mid-latitudes, with density non-decreasing with depth (for the Earth's surface the "depression" at 45° is only 4.3 meters). The oblateness increases from the center to the surface.

(C and D) The isostatic equilibrium (Figure 6). With all its mountainous ranges, its continents and its oceans, the Earth is far from the hydro-static equilibrium and a much better approximation is that of the isostatic equilibrium. This idea comes from the very large difference of viscosity between the lithosphere and the lower mantle. Vertical equilibrium of the lithosphere is achieved much more easily and faster than the horizontal one.

The upwards motion of Scandinavia, after the recent melting of the quaternary glacier, is about one meter per century and corresponds to a re-equilibrium with a time scale of a few millennia only.

The theory of isostatic equilibrium requires the definition of a "level of

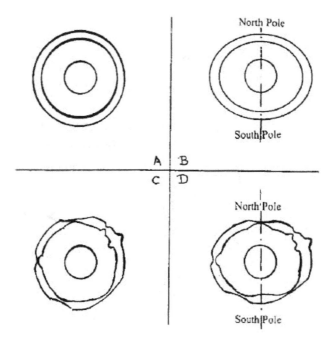

Figure 6. *The four equilibria of the Earth. (A) Hydrostatic equilibrium of a non-rotating Earth. The level surfaces are concentric spheres. (B) Hydrostatic equilibrium of a rotating Earth. The level surfaces are concentric and coaxial oblate quasi-ellipsoids (slightly depressed at mid-latitudes). (C) Isostatic equilibrium of a non-rotating Earth. The sea level is almost a sphere. (D) Isostatic equilibrium of a rotating Earth. The sea level is almost an oblate ellipsoid.*

compensation", a few hundred kilometers below the surface. Above that level only vertical equilibrium is required. We thus arrive to the usual picture with a thick lithosphere under the continents and a thinner one below the oceans (Figures 6C and 6D).

Figure 6D, isostatic equilibrium of a rotating planet, is of course the nearest to the Earth case with its main feature which is the equatorial bulge (the difference between equatorial and polar radii is 21.4 km). Figure 6C corresponds to the isostatic equilibrium of a non-rotating Earth. Its mean sea level is almost a sphere and its interest is that **the stable equilibrium position of the poles corresponds to its main axis of largest inertia** (Marchal, 1996). If the mass repartition varies, the stable position of the poles is modified accordingly.

The inertia matrix of the Earth is well known through the analysis of the nutation and precession of the polar axis and especially through the analysis

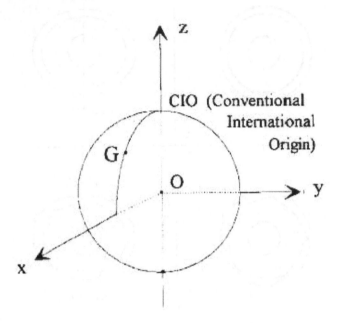

Figure 7. *The O_{xyz} rotating set of axes (International Terrestrial Reference Frame). O is the Earth center of mass. O_z passes through the Conventional International Origin (i.e. the point O of Figures 4 and 5). O_{xz} is along the Greenwich Meridian. O_{xyz} is a right and rectangular trihedron.*

of the Earth gravitational potential. For the expression of the Earth inertia matrix we will use the classical "International Terrestrial Reference Frame" (Figure 7), that is the set O_{xyz} rotating with Earth and practically fixed to the Earth crust. O is the Earth center of mass. O_z passes through the "conventional international origin" (that is the origin O of Figures 4 and 5, a few meters from the present North Pole). O_x is in the Greenwich Meridian and O_{xyz} is a right and rectangular trihedron.

The classical expression of the inertia matrix M is related to the moments and the products of inertia:

$$M = \begin{vmatrix} I(y^2 + z^2) & -I(xy) & -I(xz) \\ -I(xy) & I(x^2 + z^2) & -I(yz) \\ -I(xz) & -I(yz) & I(x^2 + y^2) \end{vmatrix} \tag{1}$$

with

$$I(x^2) = \int_M x^2 dm; \qquad I(xy) = \int_M xy\, dm; \quad \text{etc.} \tag{2}$$

where the subscript M refers to the total Earth mass and where dm is the element of mass with coordinates x, y, z in the above International Terrestrial Reference Frame O_{xyz}.

With R the Earth equatorial radius (6,378,136 m), Marchal (1996) gives the Earth inertia matrix M:

$$M = MR^2 \begin{vmatrix} U - 3.640249n4 & 1.8074n6 & 0 \\ 1.8074n6 & U - 3.577266n4 & 0 \\ 0 & 0 & U + 7.217515n4 \end{vmatrix} \tag{3}$$

The accuracy of these numbers is about $n10$ (that is 10^{-10}, according to the preliminary notice), with the only exception of the number U related to the average main moment of inertia MR^2U and only known through the precession and nutation of the polar axis:

$$U = 0.329956 \pm n6 \tag{4}$$

Fortunately this inaccuracy has no effect on the present study.

The corresponding main axes of inertia are the polar axis (largest inertia) and two equatorial axes: one with the longitude 14° 55' West (smallest inertia) and the other with the longitude 75° 5' East (mean inertia).

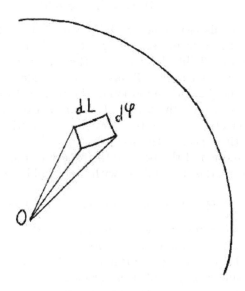

Figure 8. *A small pyramid $d\varphi$, dL with origin at the Earth center.*

The inner motions of the Earth (oceanic currents, plate tectonics, etc.) have a negligible effect on its inertia characteristics, and thus we can write the following classical equations of mechanics:

$$\mathbf{H} = \text{ Earth angular momentum} = M\omega \qquad (5)$$

with $\omega=$ the rotation vector of the International Terrestrial Reference Frame O_{xyz}.

Because of the Earth's quasi-sphericity, the directions of the angular momentum $\mathbf{H}$ and the (north) polar axis are practically the same: their angle remains always less than 2' and even less than 10" during the periods of tectonic quiescence (Marchal, 1996). Hence it is sufficient to consider the evolution of the vector $\mathbf{H}$ both with respect to a non-rotating set of axes O_{xyz} and with respect to the International Terrestrial Reference Frame O_{xyz}.

If $\mathbf{T}$ is the outer torque undergone by Earth (essentially through the attraction of the Sun and the Moon on the equatorial bulge), the evolution of the angular momentum $\mathbf{H}$ with respect to the non-rotating set of axes O_{xyz} is given by:

$$d\mathbf{H}/dt = \mathbf{T} \qquad (6)$$

However the set O_{xyz} is rotating with a rotation vector ω and thus in this set of axes the Eq. (6) becomes:

$$d\mathbf{H}/dt = \mathbf{H} \times \omega + \mathbf{T} \qquad (7)$$

Of course in this last equation the torque $\mathbf{T}$ must be expressed with its coordinates in the rotating set of axes O_{xyz}. It has then very large daily variations and a negligible daily average (because it is in the equatorial plane or very near). For that reason $\mathbf{T}$ has almost no effect on the evolution of $\mathbf{H}$ in the O_{xyz} frame (i.e. on the motion of the poles with respect to the Earth's crust). On the contrary, the torque $\mathbf{T}$ has a quasi-constant direction in the non-rotating set O_{xyz} and Eq. (6) shows a systematic motion: the precession of the poles with respect to the stars.

Hence the motion of the poles with respect to the Earth's crust is essentially governed by Eqs. (5) and (7) with a negligible torque $\mathbf{T}$:

$$\mathbf{H} = M\omega; \quad d\mathbf{H}/dt = \mathbf{H} \times \omega = (M\omega) \times \omega \qquad (8)$$

Let us assume a rotation vector ω far from the three main directions of the matrix M. The motions of the poles would be of the order of **some hundreds of kilometers per month** (with all the corresponding troubles in the ocean equilibrium), while it is zero if ω is in one of the main directions of the Earth inertia matrix M: that condition is indeed verified and it is obviously a necessary condition.

Let us consider now another repartition of the Earth mass, for instance before the melting of 50 million cubic kilometers of ice on the continents. The matrix M would be different, but the vector ω also would be different, and the matrix M is a function of the rotation vector ω because of the presence of the equatorial bulge.

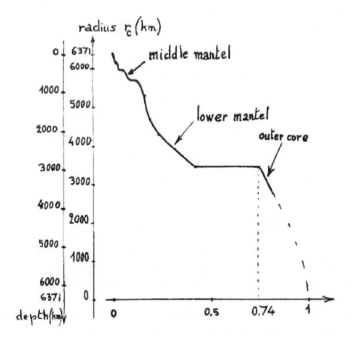

Figure 9. *The isostatic coefficient $k(r_c)$ in terms of radius r_c and the depth z of the compensation level. $r_c + z = 6371$ km.*

Let us come back to Figure 6. Figure 6D is obviously a function of the axis of rotation, but Figure 6C is not, and the corresponding inertia matrix M_C will lead to an equation similar to Eq. (8). Indeed the matrix M is practically identical to the matrix M_D of the rotating Earth in isostatic equilibrium and the principle of small modifications leads to (with M_A and M_B corresponding to the hydrostatic equilibria A and B)

$$M_C - M_D = M_A - M_B + \text{"second order"} \tag{9}$$

We will thus choose the following excellent approximation:

$$M_C = M_D + M_A - M_B = M + M_A - M_B \tag{10}$$

(for a discussion of this approximation see Marchal (1996)).
The matrix M_A has the spherical symmetry:

$$M_A = \begin{vmatrix} U_A & 0 & 0 \\ 0 & U_A & 0 \\ 0 & 0 & U_A \end{vmatrix} \tag{11}$$

The matrix M_B has the spheroidal symmetry which, with the rotation vector ω in the O_z direction, gives:

$$M_B = \begin{vmatrix} A_B & 0 & 0 \\ 0 & A_B & 0 \\ 0 & 0 & C_B \end{vmatrix} \tag{12}$$

These symmetries give always, for all rotation vectors ω:

$$(M_A\omega)\times\omega = \mathbf{0}; \qquad (M_B\omega)\times\omega = \mathbf{0} \tag{13}$$

and then with Eq. (10), in the International Terrestrial Reference Frame O_{xyz}:

$$d\mathbf{H}/dt = (M\omega)\times\omega = (M_C\omega) \times \omega \tag{14}$$

We verify that $d\mathbf{H}/dt$ is zero only when the rotation vector ω is in one of the main directions of the matrix M_C, and, since these equations are also those of an isolated rotating solid, the stable rotation is about the main axis of largest inertia.

A realistic difference $C_B - A_B$ is $1.070978\,n3\,MR^2$ (Marchal, 1996), and thus we obtain the present value of the matrix M_C for a "non-rotating Earth in isostatic equilibrium":

$$M_C = MR^2 \begin{vmatrix} U_C - 7.0322n6 & 1.8074n6 & 0 \\ 1.8074n6 & U_C - 7.339n7 & 0 \\ 0 & 0 & U_C + 7.7661n6 \end{vmatrix} \tag{15}$$

As already in Eqs. (3) and (4), the accuracy of U_C is weak, but fortunately this doesn't influence the derivative $d\mathbf{H}/dt$ and has thus no influence in this study:

$$U_C = 0.329954 \pm n6 \tag{16}$$

The two inertia matrices M and M_C have the same directions of main axes of inertia but the moments of inertia of M_C are much closer to each other:
For M:

$$A = MR^2(U - 3.645067n4) \tag{17}$$
$$B = MR^2(U - 3.572448n4)$$
$$C = MR^2(U + 7.217515n4)$$

For M_C:

$$A_C = MR^2(U_C - 7.5140n6) \qquad (18)$$
$$B_C = MR^2(U_C - 2.521n7)$$
$$C_C = MR^2(U_C + 7.7661n6)$$

Nevertheless the polar axis remains the axis of largest inertia, and thus the stability condition is satisfied. At the accuracy of this evaluation the two poles are within a kilometer from their position of stable equilibrium and of course this is not only a coincidence!

The next questions are then obviously: Is it possible to evaluate the inertia matrix M_C as it was before the melting of the last quaternary ice age? and where was the corresponding axis of main inertia, i.e. the stable equilibrium position of the poles?

3 The matrix M_c before the melting of the last ice age

Because they are independent of the Earth rotation, the matrices of the M_C type can be deduced from each other through the consideration of mass modifications. However we must also take account of the isostatic compensations: if a large glacier melt occurs in a country such as Scandinavia, this country would rise up which would reduce the effect of the mass displacement.

Let us consider first that effect.

3.1 The effect of isostasy

Let us consider the measure of a moment or a product of inertia, as $I_C(x^2)$ or $I_C(xy)$, relative to a matrix of type M_C , i.e. in a "non-rotating isostatic equilibrium configuration":

$$I_C(x^2) = \int_M x^2 dm; \qquad I_C(xy) = \int_M xy dm \qquad (19)$$

where M is the Earth mass and dm the element of mass with coordinates x, y, z in the above International Terrestrial Reference Frame O_{xyz}.

In usual geocentrical coordinates r, φ, L we have:

$$x = r \cos\varphi \cos L; y = r \cos\varphi \sin L; z = r \sin\varphi \qquad (20)$$

and, with the volumic mass ρ:

$$dm = \rho r^2 \cos\varphi dr d\varphi dL \qquad (21)$$

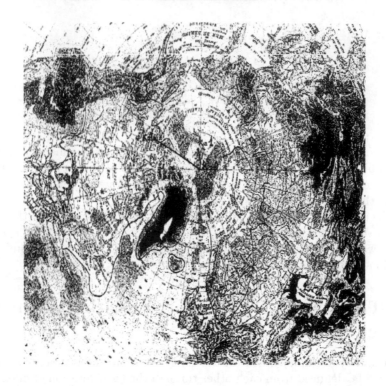

Figure 10. *The position of the North Pole in terms of the parameter $k(r_c)$ for the two hypotheses of Section 3.2 (long curve: A = 1.5 km, N = 0.5 km, W = 1.3491 km; short curve: A = N = W = 1.169 km). This astrodynamical analysis is obviously not accurate enough for a determination of the past position of the pole, but it shows that the pole could have moved hundreds of kilometers during the melting of the last ice age.*

Hence for a small pyramid $d\varphi, dL$ with origin at the Earth center (Figure 8), the contribution in a moment or a product of inertia is always proportional to:

$$\int r^2 dm; \tag{22}$$

that is proportional to

$$S(\varphi, L) = \int_0^{R + 50 \text{ km}} \rho\, r^4 dr \tag{23}$$

For instance:

$$I_C(x^2) = \int_M x^2 dm = \int_{\text{Earth}} \rho r^4 \cos^3 \varphi \cos^2 L\, dr d\varphi dL \tag{24}$$

$$= \int_{-90°}^{90°} d\varphi \int_{0°}^{360°} S(\varphi,\, L)\, \cos^3\varphi \, \cos^2 L \, dL$$

We are thus interested in the variation of that sum S when, on the spherical rectangle $d\varphi dL$, the pressure of air and water (either liquid or solid) varies from P_1 (i.e. during the glaciation) to P_2 (after the glaciation).

Table 1. *The preliminary Earth Reference Model, down to the Outer Core.*

	depth z (in km)	radius (= 6371 - z)	density	g (in m/s^2)
Ocean				
	0	6371	1.02	9.8156
	3	6368	1.02	9.8222
Lithosphere				
Upper crust	3	6368	2.60	9.8222
	15	6356	2.60	9.8331
Lower crust	15	6356	2.90	9.8332
	24.4	6346.6	2.90	9.8394
Upper mantle	24.4	6346.6	3.38	9.8394
	40	6331	3.37	9.8437
	60	6311	3.37	9.8483
	80	6291	3.37	9.8553
	115	6256	3.37	9.8664
	150	6221	3.37	9.8783
	185	6186	3.36	9.8911
	220	6151	3.35	9.9048

In the absence of isostatic motion the variation of the quantity S would be the following:

$$Raw(S_2 - S_1) = R^4(P_2 - P_1)/g \qquad (25)$$

with: $(P_2 - P_1)/g$ = variation of above air and water mass per square meter and with R = mean Earth radius = 6371 km, a little smaller than the equatorial radius R, and g = mean intensity of gravity at sea level = $g(R) = 9.8156 m/s^2$.

However we have to take account of the vertical isostatic motions. These motions are defined by a "level of compensation", at the distance r_c from the

Table 2. *The preliminary Earth Reference Model for the Transition zone (also called the "Middle mantle" or "Asthenosphere")*

depth z (in km)	radius (= 6371 - z)	density	g (in m/s^2)
220	6151	3.43	9.9048
265	6106	3.42	9.9203
310	6061	3.48	9.9361
355	6016	3.51	9.9522
400	5971	3.54	9.9686
400	5971	3.72	9.9686
450	5921	3.78	9.9790
500	5871	3.84	9.9883
550	5821	3.91	9.9965
600	5771	3.97	10.0038
635	5736	3.98	10.0038
670	5701	3.99	10.0143

center of Earth and by a uniform pressure P on that level. Below that level, the viscosity is much smaller and horizontal motions are easy.

Since the depth of the level of compensation is unknown, we will do the analysis for all possible levels.

If $g(r)$ is the gravity at the radius r (see Table 1), the pressure P at the level of r_c , the latitude φ and the longitude L is given by:

$$P(r_c, \varphi, L) = \int_{r_c}^{surface} \rho(r, \varphi, L)g(r)dr \qquad (26)$$

We will consider that the compressibility of the ocean, the crust and the mantle is negligible, which is practically true especially for the small variations of altitude considered.

If the vertical displacement of the crust at sea level is $h(\varphi, L)$ during the melting of the last glaciation, this displacement will be $R^2 h(\varphi, L)/r^2$ below, down to the level of compensation, because of the conservation of volume. Of course if $(P_2 - P_1)$ is positive, then h is negative and conversely.

With integrations on a vertical from r_c to the surface of the Earth's crust, the isostasy will give:

$$P = P_1 + \int_{r_c}^{\text{Surface}} \rho_1(r, \varphi, L)g(r)dr = P_2 + \int_{r_c}^{\text{Surface}} \rho_2(r, \varphi, L)g(r)dr \quad (27)$$

However,

$$\rho_1(r, \varphi, L) = \rho_2[(r + R^2 h/r^2), \varphi, L)] \quad (28)$$

and

$$d(r + R^2 h/r^2) = dr[1 - (2R^2 h/r^3)] \quad (29)$$

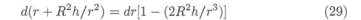

Dernier maximum glaciaire

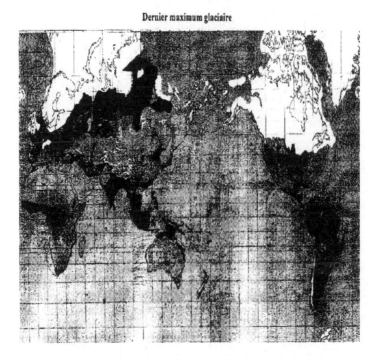

Figure 11. *Mercator map of the last ice age maximum (circa 19,000 BC).*

We will neglect the variation of g(r) between the very close levels r and $(r + R^2 h/r^2)$. These variations are very small, sometimes negative and sometimes positive (see Table 1) and furthermore they are functions of the modification of the mass repartition: the isostasy decreases them.

In these conditions, Eqs. (27) to (29) lead to the following, with subscript c for quantities at the compensation level:

$$P_2 - P_1 = hR^2[(2 \int_{r_c}^{\text{Surface}} \rho(r)g(r)dr/r^3) - (\rho_c g_c/r_c^2)] \quad (30)$$

Table 3. *The preliminary Earth Reference Model for the Lower mantle*

depth z (in km)	radius (= 6371 - z)	density	g (in m/s^2)
670	5701	4.38	10.0143
721	5650	4.41	10.0063
771	5600	4.41	9.9985
871	5500	4.50	9.9836
971	5400	4.56	9.9698
1071	5300	4.62	9.9573
1171	5200	4.67	9.9467
1271	5100	4.73	9.9383
1371	5000	4.78	9.9326
1471	4900	4.84	9.9301
1571	4800	4.89	9.9314
1671	4700	4.95	9.9369
1771	4600	5.00	9.9474
1871	4500	5.05	9.9635
1971	4400	5.10	9.9859
2071	4300	5.15	10.0156
2171	4200	5.20	10.0535
2271	4100	5.25	10.0535
2371	4000	5.30	10.1580
2471	3900	5.35	10.2272
2571	3800	5.40	10.3095
2671	3700	5.45	10.4066
2741	3630	5.49	10.4844
2771	3600	5.50	10.5204
2871	3500	5.55	10.6532
2891	3480	5.56	10.6823

$$(S_2 - S_1) = [R^4(P_2 - P_1)/g(R)] + hR^2[r_c^2\rho_c + 2(\int_{r_c}^{\text{Surface}} \rho r dr)] \qquad (31)$$

Hence if, as in Eq. (25), we call '$Raw(S_2 - S_1)$' the quantity $[R^4(P_2 - P_1)/g(R)]$

Table 4. *The preliminary Earth Reference Model for the Outer Core*

depth z (in km)	radius (= 6371 - z)	density	g (in m/s^2)
2891	3480	9.90	10.6823

that leads to mass and inertia modifications in the absence of isostatic compensation, we observe that the isostatic compensation reduces $(S_2 - S_1)$ by a factor $k(r_c)$ that is:

$$(S_2 - S_1) = k(r_c).\text{"}Raw(S_2 - S_1)\text{"} \tag{32}$$

$$k(r_c) = 1 + \{g(R)[r_c^2\rho_c + 2(\int_{r_c}^{Surface} \rho r dr)]/R^4[$$

$$(2\int_{r_c}^{Surface} \rho(r)g(r)dr/r^3) - (\rho_c g_c/r_c^2)]\} \tag{33}$$

The function $k(r_c)$ is given in Figure 9. It increases from zero to one when the depth of the compensation level increases from 0 to R. The main point is that k is only a function of r_c. It has the same value all over the surface of the Earth and this property is very useful for the determination of the inertia matrix M_C during the last ice age. Of course large values of k require a very deep compensation level, but it is not at all impossible that the influence of isostasy goes down to the outer core, that is hot and liquid and thus probably with a much smaller viscosity than the mantle. Also notice in Table 1, the big discontinuity of density at the boundary of these two main Earth zones.

3.2 Estimation of the Earth inertia matrix M_{C1} during the last ice age

Let us recall that $[Raw(S_2 - S_1)/R4]$ is a mass density on the surface of the Earth. We will measure it in kg/m^2. It will be the variation of mass of air and water (liquid and solid) on a given square meter during the melting of the last ice age. We will first consider the raw modification, i.e., without isostasy. Then we will take account of the isostatic factor $k(r_c)$ of the previous section.

The melting induces a rise of sea level of 120 meters (on average). Out of the two polar caps and the local glaciers, the variation $[Raw(S_2 - S_1)/R4]$ is positive or zero:

- It is zero on the continents and the islands.

- It represents 120 m of sea water on high seas (i.e. 122,400 kg/m^2, with a sea density of 1.02).

- In shallow seas, less than 120 m. The variation represents the depth of the sea.

In the polar caps and the local glaciers, the variation is more complex:

- In high seas, the melting has no effect and the variation is positive. It represents a 120-m rise of sea level. That rise is indeed given by the melting of continental glaciers. On the continents and the islands we will assume a negative variation corresponding to the melting of about one or two kilometers of ice.

- In shallow seas the variation is intermediate; we will also assume the melting of about one or two kilometers of ice. However 'shallow' has here a wider meaning: in a sea that is 1-km deep, a 2-km thick ice field lies on the sea bottom and has thus the properties of a continental ice field. The ice is 'efficient' down to almost the sea level.

The map of the North polar cap is given in Figures 2 and 3, and we will assume that Antarctica was entirely under the South polar cap. The global variation of mass is zero. That last condition gives the order of magnitude of the ice thickness, but of course the accuracy of the result will be weak. It seems obvious that the polar caps were thicker near the poles than along their limits and we cannot hope for accuracy better than 10 or 20%.

Notice that an ice polar cap melts essentially at its periphery, where it is warmer, and not at its top, where it is cooler. Hence we will consider that the thickness of ice in the present remaining ice polar cap (Greenland, Antarctica) is essentially as it was during the last ice age. Nevertheless, we must consider that in West Antarctica a large quantity of ice has melted. Indeed logs of ice extracted in East Antarctica enable the study of the climate back 800,000 years, while the same logs extracted in West Antarctica never go beyond the last ice minimum, 6,000 years ago, and the present thickness of ice is generally much smaller in West Antarctica.

There are then three essential zones of efficient ice melting: The Arctic (with Scandinavia and Northwestern Siberia), the North American continent and the West Antarctica.

In these conditions various hypotheses are still possible. For the sake of simplicity we will:

(A) Consider the mass modification in the direction opposite to that of the time: from the present state to the past one, hence the sea level went down.

(B) Consider the mass modification of a global decrease of 120 m of the sea level, and not only an oceanic one: the corresponding inertia modification is easy to compute. But of course the corresponding continental mass modifications are then +120 m of sea water on the continents, the complement in shallow sea, etc.

(C) Define the numbers without dimension J(ab), as the following differences:

$$MR^2 J(ab) = I_1(ab) - I_2(ab) \tag{34}$$

with, as in Eq. (3) and Eqs. (15)-(18): M = mass of Earth = 5.9737 p24 kg; R = Earth equatorial radius = 6,378,136 m and $a, b = r, x, y$ or z, the coordinates of the International Terrestrial Reference Frame of Figure 7.

$I_1(ab)$ = moment of inertia (ab) during the last ice age.

$I_2(ab)$ = present moment (or product) of inertia (ab) for matrices of type M_C and without isostatic effects.

The quantities $J(ab)$ will be decomposed into their different parts given by the global decrease of 120 m, by the different continents, etc. A global decrease of 120 m of sea level gives the following:

$$J(x^2) = J(y^2) = J(z^2) = -3.4759n6; J(x, y) = J(x, z) = J(y, z) = 0 \quad (35)$$

For the different continents, and the surrounding shallow seas, the +120 m of sea water will give results as in Table 5.

If we add to these subtotals the numbers given in Eq. 35 we will obtain all the effects independent of the melting of the efficient ice:

$J(x^2)$	$J(y^2)$	$J(z^2)$	J(xy)	J(xz)	J(yz)
-2.6129 n6	-2.3573 n6	-2.0202 n6	-1.08 n8	1.666 n7	1.585 n7

Let us now consider the effects given by the melting of the efficient ice (the continental ice and the ice in shallow seas).

For efficient ice layers 1-km thick we obtain the following in the three main zones:

	$J(x^2)$	$J(y^2)$	$J(z^2)$	J(xy)	J(xz)	J(yz)
Arctic and Northwestern Eurasia	1.598 n7	2.039 n7	2.9707 n6	3.81 n8	2.497 n7	4.326 n7
North America	4.75 n8	4.610 n7	1.3545 n6	-2.37 n8	4.1 n 9	-7.562 n7
West Antarctica	1.24 n8	2.17 n8	7.496 n7	8.1 n9	5.88 n8	1.018 n7

Notice that all these numbers that seem so accurate are only evaluations; errors of the order of 5 or 10% are certainly numerous, but it is not necessary to add new errors through the truncation of digits.

Let us call A, N and W the average thickness (in kilometers) of these three main efficient ice zones (A: Arctic and Eurasia, N: North America, W: West Antarctica). A first evaluation of these thicknesses is obtained with the condition of zero global mass variation, which is also a zero variation of $I(r^2)$, that is $I(x^2 + y^2 + z^2)$:

$$\begin{aligned} 0 &= Total J(x^2 + y^2 + z^2) = (3.3344n6)A + (1.863n6)N \\ &+ (7.837n7)W - 6.9904n6 \end{aligned} \quad (36)$$

Table 5. J values for a rise of 120 m in sea water level.

	$J(x^2)$	$J(y^2)$	$J(z^2)$	$J(xy)$	$J(xz)$	$J(yz)$
South America	9.138 n8	2.4891 n7	2.686 n8	-1.3642 n7	-3.348 n8	5.799 n8
Africa and Madagascar	4.9617 n7	8.420 n8	4.120 n8	1.4665 n7	5.090 n8	1.654 n8
Australia, New Guinea, New Zealand	8.584 n8	6.567 n8	3.480 n8	-7.176 n8	5.100 n8	-4.434 n8
Antarctica	1.106 n8	2.447 n8	3.0792 n7	0	0	-6.310 n8
Asia	7.451 n8	5.1398 n7	3.4989 n7	0	0	3.3790 n7
Europe	7.589 n8	1.563 n8	1.5381 n7	3.004 n8	1.0095 n7	4.106 n8
Caribbean and North America*	1.837 n8	1.8790 n7	2.7057 n7	2.039 n8	-2.531 n8	-1.9149 n7
Greenland and Arctic**	9.77 n9	7.88 n9	2.7061 n7	3.4 n10	2.264 n8	3.99 n9
Subtotals	8.630 n7	1.1486 n6	1.4557 n6	-1.08 n8	1.666 n7	1.585 n7

* with the Canadian arctic archipelago

** without the Canadian arctic archipelago

If, for instance, we assume $A = N = W$, Eq. (36) gives an average thickness of 1.169 km. However, there is no reason for this simple assumption and we have more information. Let us look for instance if it is possible, and reasonable, to have had a North Pole in the vicinity of the point presented in Figure 3 near the latitude 75° North and the longitude 60° West: the corresponding M_C matrix must have the main axis of largest inertia in that direction.

Let us recall the general expression of inertia matrices given in Eq. (1), and also the moments and products of inertia of the matrix M_C which is of interest, as given by Eq. (15) and the differences related to Eq. (34), the above expressions of $J(ab)$ and the isostatic reduction factor $k(r_c)$ given in Eq. (33):

$$I(y^2 + z^2) = MR^2\{U_C - 7.0322n6 + k[A \times 3.1746n6$$
$$+N \times 1.8155n6 + W \times 7.713n7 - 4.3775n6]\}$$
$$I(x^2 + z^2) = MR^2\{U_C - 7.339n7 + k[A \times 3.1305n6$$
$$+N \times 1.4020n6 + W \times 7.620n7 - 4.6331n6]\}$$
$$I(x^2 + y^2) = MR^2\{U_C + 7.7661n6 + k[A \times 3.637n7$$
$$+N \times 5.085n7 + W \times 3.41n8 - 4.9702n6]\}$$
$$I(xy) = MR^2\{-1.8074n6 + k[A \times 3.814n8 - N \times 2.37n8$$
$$+W \times 8.1n9 - 1.08n8]\}$$
$$I(xz) = MR^2\{k[A \times 2.497n7 + N \times 4.1n9$$
$$+W \times 5.88n8 + 1.666n7]\}$$
$$I(yz) = MR^2\{k[A \times 4.326n7 - N \times 7.562n7$$
$$+W \times 1.018n7 + 1.585n7]\}$$

$$(37)$$

There remain of course plenty of possibilities, but most of them lead to large pole displacements.

In order to get a taste of these possibilities, let us consider two very different cases: (1) The spectacular case $A = 1.5km, N = 0.5km$ and, with Eq. (37), $W = 1.3491km$. (2) The average, but improbable case $A = N = W = 1.169km$. The polar equilibrium positions are given in Figure 10 in terms of the isostatic reduction factor $k(r_c)$.

We can thus conclude the three following points:

1. Our present knowledge of the past Earth inertia matrix is clearly insufficient. A better analysis would be very useful with especially a more detailed knowledge of the rise of sea level. We have considered a uniform and averaged value of 120 m, but the rise goes up to 180 m in some large parts of the Indian Ocean and this difference can be crucial.

Notice that a difference of 30 meters between, for instance, the Indian Ocean and the Northern Pacific has more influence than a one kilometer difference between the thicknesses A, N and W.

2. Nevertheless, we have already reached an essential result: the melting

of the last ice age can easily have moved the polar equilibrium position by hundreds of kilometers.

3. Another important result appears: the Earth polar moment of inertia was smaller (which is natural since the water was nearer to the poles) and the difference is not at all negligible: it is about 4 n6 $MR^2 \times k(r_C)$, for any possible values of the thicknesses A, N and W.

We will see in Section 5 the corresponding disastrous consequences.

4 Are such fast and large moves of the poles possible?

We are today well accustomed with the theory of tectonic plates. The continents have a slow relative motion that can, over millions of years, modify completely the figure of the Earth. However, these motions are really very slow, usually a few centimeters per year, at most 30 or 50 cm per year. With such a velocity, the poles would never move about 1,000 km in a few millennia.

However, the motion of a pole is completely different from the relative motion of two continents. We can consider that the poles are almost fixed and that the Earth's crust falls over in one piece which, because of the Earth's quasi-sphericity, is much easier than the relative motion of two continents.

The largest obstacle to this motion is the equatorial bulge. That bulge is obviously a consequence of the Earth's rotation and it exists because the Earth is slightly viscous, which results also in its deformation. The post glacial rebound of Scandinavia: 1-m rise per century after the melting of the huge quaternary glacier gives us an order of magnitude of the viscosity of the Earth's crust: about the p21 poises, and the half-period of such motion is a few millennia. However, the half-period of viscous motions is inversely proportional to the size of the zone of interest and thus the half-period of the re-equilibrium of the equatorial bulge after a displacement of the poles is about a few centuries only.

If we consider the relatively slow rise of sea level in the past millennia (Figure 1), we can understand that the poles have followed very nearly their position of stable equilibrium, and that they have been practically fixed for the past 5,000 or 6,000 years.

5 The age of natural disasters

Our Cro-Magnon ancestors were already widely spread on Earth from at least 30,000 years ago and they lived through the Würm and Wisconsin last glaciations when, of course, the climates were very different from today.

Let's consider now the melting of these last huge quaternary glaciers. There were a lot of catastrophes: climatic upheaval, rise of sea level – which can sometimes be very fast when a depression is filled - but also big eruptions, major earthquakes and giant tidal waves. Indeed, as noted at the end of Section 3.2, we must consider that the melting has increased the polar inertia moment by about $4\ n6\ MR^2\ \times k(r_C)$, which, even if the factor k is as small as 0.1, is already a proportion of 1.2 parts per one million. Of course this proportion may seem extremely tiny, but an isolated rotating body that increases its main moment of inertia also loses a part of its kinetic energy of rotation in the same proportion.

The Earth kinetic energy of rotation is 2.13 p29 joules; 1.2 millionth of this quantity is already 2.5 p23 joules. This value represents *the energy of about 20,000 major earthquakes.* Of course a very small part of this energy is lost into the re-equilibrium of the equatorial bulge (less than 1%), but its large majority must be dissipated into phenomena that have all chances to be terrible.

6 Conclusion

This astrodynamic study of the Earth's rotation during the melting of the last quaternary ice age gives only partial results: the melting of the huge quaternary glaciers could easily have moved the poles by hundreds of kilometers and even more, but the accuracy of the reconstitution of the Earth inertia matrix remains insufficient to confirm that the pole was really near the present Baffin Sea as suspected in Figure 3. It would be of interest to do the same analysis again, with a more accurate map of the rise of sea level than the above hypothesis of a uniform rise of 120 meters. That value of 120 meters represents the average value, but we already know that the rise reached 180 meters in some large places.

The most interesting and certain result is the increase of a few millionths of the Earth polar moment of inertia, with the corresponding decrease of the Earth energy of rotation in the same proportion. That small decrease represents nevertheless a fantastic amount of energy and it passes into innumerable very energetic phenomena (eruptions, earthquakes, tidal waves, etc.), which with the climatic upheaval and the rise of sea level have certainly made the period of the melting - essentially from 13,000 BC to 8,000 BC - an age of natural disasters.

It is of course very impressive to notice that in all civilizations—West, Middle East, Far East, Pre-Columbian America, Africa, Australia, Oceania—the myths tell us stories of ancient catastrophes similar to the Great Flood and the Lost Paradise.

References

Balmino, G., Reigber, C. and Moynot, B. (1978), Le modèle de potentiel gravitationnel terrestre GRIM 2, *Annales de Géophysique* **34** (2), 55-78.

Lerch , F.J. et al. (1979), Gravity model improvement using GEOS-3 (GEM-9 and 10), *Journal of Geophysical Research* **84**, 3897-3915.

Marsh, J.G. et al. (1990), The GEM-T2 gravitational model, *Journal of Geophysical Research* **95**, 22043-22071.

Reigber, C. et al. (1993), Grim 4-C1-C2p: Combination solutions of the global Earth gravity field, *Surveys in Geophysics* **14**, 381-393.

Marchal, C. (1996), Earth's polar displacements of large amplitude: a possible mechanism, *Bulletin du Muséum National d'Histoire Naturelle* **18**, 517-554.

Marchal, C. (1968), Figures d'équilibre séculairement stables des masses fluides hétérogènes en rotation, *Bulletin Astronomique* **3**, 341-360.

Marchal, C. (1997), Errata of "Earth's polar displacements of large amplitude: a possible mechanism", *Geodiversitas* **19**, 159.

US. National report to the International Union of Geodesy and Geophysics 1991-1994 (1995), Sea level rise in the recent millennia, *Also: American Geophysical Union. Review of Geophysics* **39**.

List of Participants

David Anderson	Keele University, UK dra@astro.keele.ac.uk
Benjamin Ayliffe	University of Exeter, UK ayliffe@astro.ex.ac.uk
Mariangela Bonavita	Observatory of Padova, Italy mariangela.bonavita@oapd.inaf.it
Mark Booth	Institute of Astronomy, Cambridge, UK mbooth@ast.cam.ac.uk
Adrian Colucci	Universidade de Sao Paulo, Brazil adrian@astro.iag.usp.br
Patricia Cruz	Universidade de Sao Paulo, Brazil pcruz@astro.iag.usp.br
Sara Di Ruzza	Universit "La Sapienza", Rome, Italy diruzza@mat.uniroma1.it
Linda Dimare	Universit "La Sapienza", Rome, Italy dimare@mat.uniroma1.it
Diana Dragomir	McGill University, Ontario, Canada dragomir@hep.physics.mcgill.ca
Fawzy Elsabaa	Faculty of Education, Ain Shams University, Egypt elsabaa@hotmail.com
Francesca Faedi	University of Leicester, UK ff20@star.le.ac.uk
Nikolaos Georgakarakos	University of Edinburgh, UK georgakarakos@hotmail.com
Mathew Giguere	San Francisco State University, USA magickair@gmail.com
Cristian Giuppone	Observatorio Astronomico de Cordoba, Argentina cristian.giuppone@gmail.com
Alex Golovin	Kyiv National Taras Shevchenko University, Ukraine astronom_2003@mail.ru
Christianne Helling	University of St. Andrews, UK c.helling@st-andrews.ac.uk

Krzysztof Helminiak   Polish Academy of Sciences, Poland
xysiek@ncac.torun.pl

Claudia Hernandez Mena   Universidad Autonoma del Estado de Morelos, Mexico
cmena@ce.fis.unam.mx

Howard Isaacson   San Francisco State University, USA
isaacson@stars.sfsu.edu

Anne-Sophie Libert   FUNDP - Departement de Mathématique, Belgium
anne-sophie.libert@fundp.ac.be

Elke Lohinger   University of Vienna, Austria
lohinger@astro.univie.ac.at

Zoltan Mako   Sapientia University, Romania
makozoltan@sapientia.siculorum.ro

Cezary Migaszewski   Nicolaus Copernicus University, Poland
Cezary.Migaszewski@astri.uni.torun.pl

Althea Moorhead   University of Michigan, USA
altheam@umich.edu

Ian Morison   University of Manchester, UK
Ian.Morison@manchester.ac.uk

Jennifer Noble   Edinburgh University, UK
s0233201@sms.ed.ac.uk

Matthew Payne   Institute of Astronomy, Cambridge, UK
mpayne@ast.cam.ac.uk

Ettore Perozzi   Telespazio, Rome, Italy
ettore.perozzi@telespazio.com

Neil Philips   University of Edinburgh, UK
n.m.phillips@sms.ed.ac.uk

Zsolt Sandor   Eötvös University, Hungary
Zs.Sandor@astro.elte.hu

Muhammad Shoaib   University Technologi Petronas, Malaysia
safridi@gmail.com

Anoop Sivasankaran   Glasgow Caledonian University, UK
asi6@gcal.ac.uk

Alexis Smith   University of St. Andrews, UK
amss@st-andrews.ac.uk

Paul Steele   University of Leciester, UK
prs15@star.le.ac.uk

Piotr Sybilski   Nicolaus Copernicus University, Poland
Piotr.Sybilski@astri.uni.torun.pl

Ferenc Szenkovits   Babes-Bolyai University, Cluj-Napoca, Romania
fszenko@math.ubbcluj.ro

Karla Torres   INPE, Brazil
karlchen79@gmail.com

Index

v Andromedae, 114

absorption cross-section, 59
absorption optical thickness, 59
aerosol, 58
angular Einstein radius, 19
angular-momentum deficit, 219
anomaly detector, 44
Antarctica, 260, 278, 279
apsidal corotation resonance, 216
Arnold cat, 186, 187
Arnold diffusion, 201
Arnold web, 206
Arnold's model, 200
asteroids, 92, 93
Asthenosphere, 273
astrocentric elements, 231–233, 235
astrocentric equations of motion, 223
asymmetric stellar jets, 110

bilinear relation, 178
Box-Least-Squares (BLS) algorithm, 9

caustic, 26, 28, 30
chaos, 185
chaotic dynamics, 185, 190, 195, 196
column number density, 59
comets, 92, 93, 95, 96, 99
conformal mapping, 172
convergent migration, 113
Copernican principle, 89
core accretion model, 43, 84, 86, 87, 101
CoRoT, 116
correlated errors, 6
Cro-Magnon, 259, 282

debris discs, 92–94, 96, 98, 99, 103
degree of polarization, 52
Delaunay elements, 229
depolarization factor, 59
detected planets, 38
detection efficiency, 31, 42
disk fragmentation model, 101

disk morphology, 102

Earth, 259, 260
Earth gravitational potential, 266
Earth gravitational potential, 262
Earth inertia matrix, 259, 266–268, 277, 281, 283
Earth kinetic energy of rotation, 283
Earth polar moment of inertia, 259, 282, 283
Earth poles, 259
Earth's crust, 268, 275, 282
Earth's rotation, 259, 263, 283
Earth-mass planet, 44
eccentric anomaly, 171
eccentricity excitation, 107
Eddington, 18
Einstein, 17
Einstein circle, 20
Einstein radius, 19
energy equation, 173
equatorial bulge, 263, 282
extinction cross-section, 59
extinction optical thickness, 58
extra-solar planets, 3

Fast Lyapunov Indicator (FLI), 202
fibration, 178
flux, 52

gas-to-dust mass ratio, 82, 84, 85
geometric albedo, 57
GJ 876, 113
GOME, 64
gravitational instability, 103
gravitational radius, 18
Great Flood, 283

habitable zone, 42
harmonic oscillator, 174, 181
HATnet, 4
HD 128311, 114
HD 73526, 114
hierarchical planetary systems, 111
homoclinic tangles, 196

hydrostatic equilibrium, 263–265, 269
hyperbolic saddle points, 188

ice age, 259
ice melting, 278
infrared photometry, 3
interferometric observations, 102
International Terrestrial Reference Frame, 266–268, 270, 271, 279
isolation mass, 104
isostasy, 271, 275, 277
Isostatic equilibrium, 264, 265, 269, 270

KAM theorem, 202
Kepler, 116
Kepler motion, 170
Keplerian elements, 231–234
Keplerian equation, 171
KS regularization, 175
Kuiper belt, 92–94, 96, 100

Lambertian surface, 63
Laplace-Lagrange theory, 220
late heavy bombardment, 94, 96, 98
lens equation, 18
lensing zone, 28
Levi-Civita regularization, 173
lithosphere, 264, 265, 273
Lost Paradise, 283
low-eccentricity planets, 217, 218
lower crust, 273
lower mantle, 264

MACHO, 37
Markov-chain Monte-Carlo (MCMC) methods, 12
mean-motion resonance, 214, 216, 219, 221
mean-motion resonant planetary systems, 111
Mercator projection, 261
Metropolis-Hastings rule, 14
microlensing, 35
middle mantle, 273
minimum-mass solar nebula, 84, 86
MOA, 36

North Pole, 260–263, 266

oblate satellite, 187
OGLE, 36
OGLE-2003-BLG-235/MOA-2003-BLG-53Lb, 38
OGLE-2005-BLG-071Lb, 39
OGLE-2005-BLG-390Lb, 39
oligarchic growth, 104
oligarchs, 104
optical depth, 22
orbital migration, 105
outer core, 277

periodic orbit, 195
perturbed Kepler problem, 180
perturbed linear problem, 170
phase angle, 53
PLANET, 36
planet abundance, 38, 42
planet detection, 17, 35
planet detection efficiency, 31, 42
planet-disk interactions, 110
planetary radiation, 52
planetary scattering matrix, 57
planetary signal, 30
planetesimals, 81, 82, 84–87
Poincaré's relative canonical coordinates, 227
polar equidistant projection, 261
polarization, 52
pole displacements, 281
primoridal disks, 103
proto-planetary discs, 92, 94, 98
protostellar classification scheme, 82, 84
pulsar planets, 211, 213

quaternary glacier, 260, 264, 282
quaternion algebra, 175

Raman scattering, 67
Rayleigh scattering, 62
red edge, 65
red noise, 8
reflected flux, 56
rigid body, 263
RoboNet, 37

saddle fixed point, 197
scattering cross-section, 59

scattering optical thickness, 59
secular planetary systems, 111
secular resonance, 220, 221
SIGNALMEN, 44
single scattering albedo, 61
single scattering matrix, 61
Smale horseshoe, 185
Smale horseshoe map, 190, 195
spectral energy distribution, 82, 83,
        87, 102
spheroidal symmetry, 264, 270
stable equilibrium position of the poles,
        259
standard map, 187
strong planet-planet scattering, 108
symplectic map, 187
SysRem algorithm, 6

T Tauri stars, 84
Transatlantic Exoplanet Survey (TrES),
        4
Trend-filtering algorithm (TFA), 7

upper crust, 273
upper mantle, 273

Würm, 259–262, 282
WASP Project, 4
Wisconsin, 259–262, 282

XO Project, 4

zero-point correction, 5

Milton Keynes UK
Ingram Content Group UK Ltd.
UKHW021617071024
449327UK00020BA/1090